# MADAGASCAR WILDLIFE

Hilary Bradt
Derek Schuurman
Nick Garbutt

Bradt Publications, UK
The Globe Pequot Press Inc, USA

First published in 1996 by Bradt Publications,
41 Nortoft Road, Chalfont St Peter, Bucks SL9 0LA, England
Published in the USA by The Globe Pequot Press Inc,
6 Business Park Road, PO Box 833, Old Saybrook, Connecticut 06475-0833

**British Library Cataloguing in Publication Data**
A catalogue record for this book is available from the British Library
ISBN 1 898323 40 2
**US Library of Congress Cataloging in Publication Data**

**Illustrations**
Nick Garbutt

**Cover photographs**
Front: Sportive lemur *(Lepilemur ruficaudatus)* (NG)
Back: Chameleon *(Furcifer labordi)* (MB)

**Photographs**
Josephine Andrews (JA), Anne Axel (AA), Quentin Bloxam (QB), Deri Bowen (DB),
Hilary Bradt (HB), John Buchan (JB), Marius Burger (MB), Adrian Deneys (AD),
Nigel Dennis (ND), Nick Garbutt (NG), Steve Garren (SG), Ben Gaskill (BG),
Horst Gossler (HG), Dominique Halleaux/BIOS photo library (DH/BIOS),
Caroline Harcourt (CH), Paul Hellyer (PH), Johan Hermans (JH),
Olivier Langrand/BIOS photo library (OL/BIOS), William Love (WL),
Peter Morris (PM), Mark Pidgeon (MP), Frances Pirie (FP),
Brian Rogers (BR), Gavin and Val Thomson (GT)

**Map**
Steve Munns

Designed by Ian Chatterton (01484 456156)

Printed in Hong Kong by Colorcraft Ltd

# CONTENTS

# ABOUT THIS BOOK

'**W**hat on *earth* is that?' asked a visitor to Madagascar as she stared at the leaf-tailed gecko in disbelief. 'Are we looking at *fulvus rufus* or *fulvus collaris*?' queried the zoologist as he observed some brown lemurs in Berenty. This book answers the questions of all visitors to Madagascar, whether holidaymakers or serious naturalists. Madagascar may lie a mere 400 kilometres off the coast of Africa, but it is separated from it by millions of years of evolution: nature's private sanctuary.

In compiling this book we have selected only the most interesting, appealing or beautiful animals, generally limiting our choice to those easily seen by visitors. And to help in the choice of places to visit, an introductory chapter describes the most accessible reserves and the key species likely to be seen there.

## MADAGASCAR AND MALAGASY NAMES

The adjective 'Malagasy' is generally used in preference to 'Madagascan' by both naturalists and general writers. Malagasy is also the name of the people of Madagascar.

After independence the Malagasy were naturally anxious to replace the French colonial names of their major towns. This has posed a problem for foreigners since the Malagasy language is full of long, hard-to-pronounce names, so even the best-intentioned tour operators and visitors tend to use the old names. To avoid possible confusion we have done the same in this book, with the Malagasy name in parentheses.

## ACKNOWLEDGEMENTS

The authors would like to thank the experts who generously supplied information for this book: Caroline Harcourt, Clare Hargreaves, Mike & Liz Howe, Alison Jolly, Olivier Langrand, David Lees, Gilbert Rakotoarisoa (JWPT, Madagascar), Nivo Ravelojaona, Don Reid, Ian Sinclair, Hilana Steyn and Lucienne Wilmé.

Many thanks also to the following photographers who did not make it on to page v: Josephine Andrews, Anne Axel, Deri Bowen, John Buchan, Nigel Dennis, Steve Garren, Olivier Langrand, Mark Pidgeon and Frances Pirie; they all provided crucial images, often at the last minute. A full list of photographers' names appears on page ii; their initials appear beside their photographs.

# AUTHORS AND PHOTOGRAPHERS

**Hilary Bradt** first learned about Madagascar in 1974 when she attended an illustrated talk given by a zoologist in Cape Town. Her first visit to the island was made two years later. She has led tours there since 1982 and is the author and publisher of *Guide to Madagascar* (Bradt Publications) and *Madagascar* in the World Bibliographic series published by Clio Press.

**Nick Garbutt** is a zoology graduate of Nottingham University. He has a special interest in Madagascar, making numerous visits to investigate and photograph the island's natural history and to lead wildlife tours. His photographs have been highly commended in the BBC Wildlife Photographer of the Year competition. He is currently working on *An Illustrated Guide to the Mammals of Madagascar* for Pica Press.

**Derek Schuurman** is a keen naturalist working for Unusual Destinations, a Johannesburg tour operator specialising in Madagascar. He has written numerous articles about the island. His book *Globetrotter Travel Guide to Madagascar* is published by Struik/New Holland.

**Marius Burger** is a research assistant at Eastern Cape Nature Conservation in South Africa. He leads 'herping' trips to Madagascar. His photographs have illustrated many specialist herpatological articles and books. He provided most of the *Reptiles and Frogs* chapter in this book.

**Quentin Bloxam** is curator of reptiles at the Jersey Wildlife Preservation Trust and has led numerous natural history tours to Madagascar. He is currently working on a photographic guide to Kirindy Forest.

**Ben Gaskill** first visited Madagascar in 1986 as the bat expert on a British expedition to Ankarana. He now runs a quartz export business there.

**Johan and Clare Hermans** are award-winning photographers and horticulturalists. They have a special interest in the orchids of Madagascar, and have completed a bibliography on the subject.

**William (Bill) Love** is a nature photographer, writer, lecturer and private breeder of reptiles and amphibians in Alva, Florida. His interest in everything Malagasy has led him into guiding ecotours to Madagascar and other tropical places.

**Peter Morris** is an ornithologist and tour leader currently working on *An Illustrated Guide to the Birds of Madagascar* for Pica Press.

**Brian Rogers** is a recently retired GP who also has an honours degree in zoology and has had an interest in wildlife photography for many years, his images appearing in numerous wildlife magazines (through the agency Biofotos). He has been commended in the BBC Wildlife Photographer of the Year competition.

**Gavin and Val Thomson** have made two lengthy visits to Madagascar. Their main interest is birds, but on a recent visit they managed to photograph the elusive fosa.

Madagascar pygmy kingfisher *(Ispidina madagascariensis)*

# AN INTRODUCTION TO THE WILDLIFE OF MADAGASCAR

Madagascar is truly the naturalist's promised land. Here nature seems to have withdrawn into a private sanctuary in order to work on designs which are different from those she has created elsewhere. At every step you are met by the most bizarre and wonderful forms.

*Joseph Philibert Commerson, 1771*

**M**adagascar had been enchanting naturalists long before Charles Darwin visited another isolated group of islands and began developing his theories on evolution. Unlike the Galapagos, which rose out of the sea through volcanic eruptions, Madagascar was once part of a huge land mass that covered the entire southern hemisphere. Some 200 million years ago, when dinosaurs roamed the earth, Gondwanaland, or Gondwana, began to break up into the present-day continents of Africa, Asia, South America, Antarctica and Australasia. Madagascar broke away from Africa about 165 million years ago, before the evolution of mammals.

Reptiles and invertebrates would have been the original stowaways as the new island drifted east. Later life forms such as mammals would probably have had to raft across the Mozambique Channel on clumps of vegetation. Not many made it. Absent from Madagascar are the more familiar large carnivores such as cats, dogs and bears. Absent, too, are their typical prey animals: antelope, deer and other ungulates.

Geological forces thrust up the high mountain range that runs down the centre of the island, creating dramatically different climates – wet in the east, dry in the west and south.

*Right* Granite outcrop, central highlands (near Fananarantsoa)

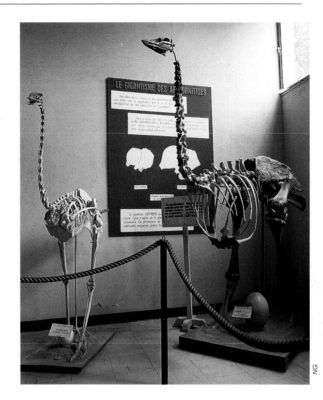

*Right* Skeleton of an elephant bird *(Aepyornis maximus)*. This huge flightless bird, probably the largest that ever lived, is shown next to the skeleton of an ostrich. Note the massive thigh bones. It is likely that the *Aepyornis* became extinct only relatively recently, around the time of the dodo. Sub-fossil egg shells and even complete eggs can still be found in the south of Madagascar.

The flora evolved according to the rainfall, and the fauna evolved to fill the multitude of niches on this mini-continent of rainforest, deciduous forest, rocky mountains and semi-desert. It is said that on Madagascar there are 200,000 different species of living things, of which 150,000 are found nowhere else on earth. In all, 83% of the wildlife is unique to the island, (that figure goes up to an astonishing 97% if birds are excluded), and much of it is startlingly different from anything seen elsewhere in the world. Truly the naturalist's promised land.

Two thousand years ago an all-conquering predator arrived in this Garden of Eden: man. A thousand years later two dozen species of large animals had been wiped out, including 15 species of lemur, some as big as gorillas, and the largest bird that ever lived, the elephant bird *(Aepyornis maximus)*. It took close on another thousand years for man to recognise his folly and start protecting instead of destroying. In 1927 the French colonial government created the first reserves. Conservation took a back seat after independence in 1960 until international interest in this treasure house of nature prompted the Malagasy government to draw up a conservation programme.

Today, Madagascar is considered one of the world's top conservation priorities. The national parks and special reserves are being developed for ecotourism, the country's best chance of earning the foreign currency needed to continue its efforts to save its priceless heritage.

# Evolution

Why is the wildlife of Madagascar so different? To find the answer, we need to understand the forces that create all living things: the forces of evolution.

Individuals of a species differ slightly from one another: these differences are the result of variations in their genes. Sometimes these differences convey an advantage that gives the individual a better chance of survival: that is to say improves its 'fitness'. It follows that such individuals are more likely to breed and, crucially, pass the beneficial genes to their offspring. This is called *natural selection*. With successive generations the advantageous genes will spread through the population.

Organisms live in environments that change constantly, and individuals with characteristics that best suit the prevailing conditions will survive at the expense of those that are not suited. Over successive generations these favourable characteristics (adaptations) will accumulate (by natural selection) and eventually cause the organism to alter. This is *evolution*.

When Madagascar became isolated from Africa, conditions on the new island were different from those on the mainland. With time, the 'founding stock' of animals and plants evolved in response to their new circumstances, to become new species (a process called *speciation*). And because Madagascar was isolated there were no diluting influences from the mainland, so the new species could develop in unique ways. However, some species on Madagascar do show some similarities to species in other parts of the world. For example, the *Mantella* frogs look and behave very much like the poison-arrow frogs from Central and South America, even though they are only distantly related. Both live similar life-styles under similar conditions, so natural selection has independently arrived at a similar solution: this is called *convergent evolution*.

This may also happen to unrelated organisms living in the same area, where they develop common characteristics (through convergent evolution) and then continue to evolve along similar lines, so that their resemblances become quite striking: this is called *parallel evolution*. For example, the true sunbirds and sunbird-asties on Madagascar.

*Above left* The green-backed Mantella frog, Madagascar.

*Above right* Poison-arrow frog, South America. Unrelated, yet so similar: an example of convergent evolution.

# WHAT'S IN A NAME?

## TECHNICAL TERMS

We have made every effort to keep the text in this book accessible. However, the use of some technical terms has been essential to explain fully certain aspects of natural history. The following definitions will unravel any mysteries.

**Nocturnal**   Animals active only under cover of darkness (at night).

**Diurnal**   Animals active only during the daylight hours.

**Cathemeral**   Animals that are active both by day and by night.

**Crepuscular**   Animals that confine their activity to the twilight hours around dusk and dawn.

**Larva**   The immature stage in an insect's life-cycle between the egg and the pupa, eg caterpillar or maggot. The pupa then undergoes complete metamorphosis into a mature adult.

**Nymph**   The immature stages of insects that resemble the adult forms and develop in gradual steps, eg crickets and grasshoppers.

**Hibernate**   Where animals lower their metabolic rate and become torpid in response to cold (normally winter). They survive on fat reserves laid down during the summer.

**Aestivate**   Where animals enter torpor in response to a dry season, also living on fat reserves.

**Endemic**   A species (or other taxonomic group) that is restricted to a particular geographic region, often due to isolation, as with islands. Both taxon and region must be defined, eg crowned lemurs are endemic to northern Madagascar, while their family, Lemuridae, is endemic to the island as a whole.

**Indigenous**   A species (or other taxonomic group) that naturally occurs in a particular region, but that also naturally occurs in other regions, eg pied crows are found throughout sub-Saharan Africa but also Madagascar and are indigenous to both areas.

**Exotic**   A species (or other taxonomic group) that has been either deliberately or accidentally introduced to a region to which it is not indigenous.

**Niche**   The role of an organism within its physical environment and with respect to the communities of other organisms that share its environment.

**Gene**   The segments of DNA contained in cells that carry the blueprints (in code form) for building all living organisms.

# CLASSIFICATION

Biologists have created a strict set of rules to identify organisms precisely. This is called classification or taxonomy and an understanding of its principles will help clarify the descriptions in this book.

The first division, called a kingdom, is very broadly defined, for instance animal, plant or fungi. Each subsequent division then becomes more precisely defined, until the organism is identified: this is called the species. Here are two *simplified* examples from Madagascar.

| Ring-tailed lemur | | Parson's chameleon | |
|---|---|---|---|
| Kingdom: | Animalia | Kingdom: | Animalia |
| Phylum: | Chordata | Phylum: | Chordata |
| Class: | Mammalia | Class: | Reptilia |
| Order: | Primates | Order: | Squamata |
| Family: | Lemuridae | Family: | Chamaeleonidae |
| Subfamily: | Lemurinae | Subfamily: | Chamaeleoninae |
| Genus: | *Lemur* | Genus: | *Calumma* |
| Species: | *catta* | Species: | *parsonii* |

The different chapters in this book broadly relate to different Classes of animals, for instance mammals (Mammalia), birds (Aves) and reptiles (Reptilia). Given in Latin, the universal language of biologists, only the genus (plural: genera) and species are quoted. This constitutes the 'scientific name' and is written in italics. The genus name always begins in upper case, while the species name is written entirely in lower case. If the genus is known, but not the species, it will be written thus: *Lemur* sp., or if the description includes several members of the same genus (congeners), it will be written: *Lemur* spp.

The scientific name often gives useful information about the animal's appearance, preferred habitat, where it lives or who discovered it. Take for example, the red-bellied lemur *(Eulemur rubriventer)*: *ruber* means red, while *venter* means belly. If the species name is *occidentalis*, then the animal is from western Madagascar, whilst you can deduce that the golden-crowned sifaka *(Propithecus tattersalli)* was named after the primatologist Ian Tattersall.

Furthermore some species are divided into subspecies or races. Often these groups have been separated for long periods, perhaps by geographical barriers, and evolution has changed, say, their colour (although they could interbreed to produce hybrids were their populations to come together again). The subspecies name is written after the species name. For example in the brown lemur, the benchmark or 'nominate' race is *Eulemur fulvus fulvus*, plus five other subspecies, the white-fronted brown lemur *(Eulemur fulvus albifrons)*, red-fronted brown lemur *(Eulemur fulvus rufus)*, Sanford's brown lemur *(Eulemur fulvus sanfordi)*, white-collared brown lemur *(Eulemur fulvus albocollaris)* and collared brown lemur *(Eulemur fulvus collaris)*.

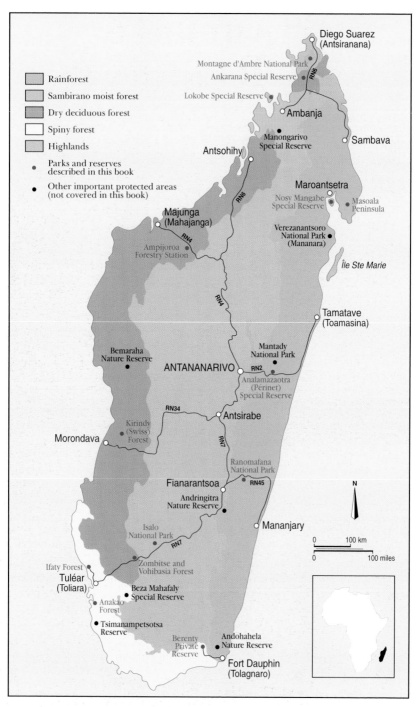

Diego Suarez
(Antsiranana)

Montagne d'Ambre National Park
Ankarana Special Reserve
RN6

Lokobe Special Reserve

Ambanja

Rainforest
Sambirano moist forest
Dry deciduous forest
Spiny forest
Highlands
Parks and reserves
described in this book
Other important protected areas
(not covered in this book)

Manongarivo
Special Reserve

Sambava

Antsohihy

Maroantsetra

Nosy Mangabe
Special Reserve

Masoala
Peninsula

RN6

Majunga
(Mahajanga)

RN4

Ampijoroa
Forestry Station

Verezanantsoro
National Park
(Mananara)

Île Ste Marie

RN4

Tamatave
(Toamasina)

Bemaraha
Nature Reserve

Mantady
National Park

ANTANANARIVO    RN2

Analamazaotra
(Périnet)
Special Reserve

RN34

Antsirabe

Morondava

Kirindy
(Swiss)
Forest

RN7

Ranomafana
National Park

Fianarantsoa    RN45

Andringitra
Nature Reserve

N

Mananjary

Isalo
National Park

RN7

0          100 km

0          100 miles

Ifaty Forest

Zombitse and
Vohibasia Forest

Tuléar
(Toliara)

Anakao
Forest

Beza Mahafaly
Special Reserve

Tsimanampetsotsa
Reserve

Berenty
Private
Reserve

Andohahela
Nature Reserve

Fort Dauphin
(Tolagnaro)

# HABITATS
# AND RESERVES

Ranomafana National Park

# RAINFOREST

The eastern rainforest band is a treasure chest for naturalists as it supports the greatest number of animal and plant species in Madagascar. To visitors more familiar with temperate woodlands, where one or two species like beech or maple dominate, the diversity of plants in a small area of tropical rainforest is bewildering. In Madagascar botanists are still classifying the flora: palm specialists have recently discovered many new species, and orchids, for which the island is famous, are also receiving considerable attention.

Evergreen rainforest in Madagascar has different characteristics depending on altitude and rainfall. Most of the protected areas described here fall into the two categories outlined below.

**Lowland rainforest** lies below 800m and is drenched with an average of 3,500mm of rain per year. It is similar to rainforest growing in other tropical regions, with large trees supported by buttress roots, smaller trees with stilt-like aerial roots, saplings, lianas, epiphytes and ferns. Compared with other continents, however, the trees are closer together and the canopy lower, and there are fewer tall trees poking their crowns above the rest. Species of *Impatiens* are common here.

**Montane rainforest** is typical of key reserves like Analamazaotra (Périnet), Ranomafana, and Montagne d'Ambre (Amber Mountain). It is found between 800m and about 1,300m, above which the name high altitude montane forest is used. Tree ferns are a feature of montane forests and bamboo is common, as are species of *Kalanchoe*. Compared with lowland rainforest, leaves are smaller and tougher, mosses and epiphytes are more abundant and shrubby undergrowth flourishes as more light reaches the forest floor.

The majority of Madagascar's rainforests are concentrated in a band, known as the Madagascar Sylva, which extends from around Iharana (Vohemar) in the north all the way down to Tolagnaro (Fort Dauphin) in the south. Once continuous, this band of forest has been severely fragmented by the timber industry, and – more importantly – through the slash and burn agricultural practices of the land-hungry rural people.

Other isolated patches of montane forest occur towards the island's northern tip around Montagne d'Ambre. A further area of moist forest in the northwest, the Sambirano Domain, constitutes a transition between eastern rainforests and western deciduous forests. The main block is centred around Manangorivo and part of the Tsarantanana Massif, but it also extends to the coast and includes the forests on Nosy Be and some of its surrounding islands.

# LOKOBE SPECIAL RESERVE

A visit to Lokobe is one of the most popular excursions on the holiday island of Nosy Be. The Black Lemur Forest Project is working with local communities on conservation, and visits by tourists are likely to be extended in the reserve to raise awareness of the problems facing the lemurs here. Lokobe offers a good introduction to the ecology of a coastal rainforest, and can be combined with the marine reserve island of Nosy Tanikely, where a small patch of forest provides roosting trees for fruit bats, and Nosy Komba (see opposite).

## Habitat and Terrain
Lokobe is the last remaining stand of lowland rainforest left on Nosy Be and covers an area of 740ha. It is part of the Sambirano Domain, an enclave of moist forest that is characterised by a distinct dry spell from June to August. The forest appears very similar to the more 'typical' rainforests of the east, but there are a number of locally endemic plant species. Where large trees remain the canopy reaches 30–35m.

## Key Species
**Mammals** Black lemur, grey-backed (Nosy Be) sportive lemur, brown mouse lemur, greater hedgehog tenrec.

**Birds** Hook-billed vanga, cuckoo roller, Madagascar blue pigeon, Madagascar pygmy kingfisher, Madagascar long-eared owl, white-throated rail, Madagascar paradise flycatcher.

**Reptiles and frogs** Panther chameleon, Boettger's chameleon, Madagascar ground boa. Less easily seen are at least four stump-tailed chameleons including *Brookesia minima*, two leaf-tailed geckos, *Uroplatus henkeli* and *U. ebenaui*, and *Mantella betsileo*, a pretty little orange-backed frog.

## Visitor Information
**Location** A peninsula at the southeast corner of Nosy Be, 5km from Hell-Ville.

**Access** At present the reserve is reached by canoe (*pirogue*) as part of an organised excursion. Easier access by vehicle may be possible in the future. Unescorted visits are not permitted. Within the reserve there are good trails.

**Best months to visit** November to March to see panther chameleons in their breeding colours. Cooler and drier between May and September.

**Accommodation** The reserve is easily reached from Nosy Be's many hotels.

**Grading** Moderate. The heat and steep trails are tiring.

**Recommendations** Do not expect to see pristine rainforest, nor to deviate from the standard tour. You will come away with a better understanding of how the local people use the forest and its products, and you will see plenty of wildlife.

HB

HG

*Above* Sandy beach at Lokobe

*Left* Nosy Komba, a small island off Nosy Be, is *the* place to get that 'lemur on the shoulder' experience. Black lemurs have always lived unmolested by the villagers who consider them sacred, and for years visitors have fed them bananas (a practice frowned on in other reserves, but the villagers here live off the income generated by their tame black lemurs).

# MONTAGNE D'AMBRE (AMBER MOUNTAIN) NATIONAL PARK

Montagne d'Ambre is a green oasis: Diego Suarez (Antsiranana) receives only about 900mm of rain per year whilst the park is drenched with an average of 3,585mm. This is perhaps Madagascar's most rewarding reserve for the average visitor. It is easy to get to, has a good trail system, labelled trees and points of interest (some in English), and is very beautiful. Lemurs and many bird species are usually seen.

## Habitat and Terrain
Montagne d'Ambre is an isolated patch of montane rainforest covering an area of 18,200ha and lying at altitudes between 850m and 1,475m. It derives its name from the resin that oozes from some of its trees, a few of which reach 40m. The park is notable for its bird's nest ferns, tree ferns, orchids, mosses and lianas. Two waterfalls form the focal points, and there are crater lakes and viewpoints over the forest and surrounding area.

## Key Species
**Mammals** Sanford's brown lemur, crowned lemur, northern sportive lemur, Amber Mountain fork-marked lemur, brown mouse lemur, northern ring-tailed mongoose and fosa. The rare falanouc has been seen here.

**Birds** Madagascar crested ibis, Malagasy kingfisher, Madagascar blue pigeon, forest rock-thrush, Madagascar magpie-robin, broad-billed roller, cuckoo-roller, pitta-like ground roller, Madagascar paradise flycatcher, souimanga sunbird, Madagascar white-throated rail, white-throated oxylabes, spectacled greenbul, hook-billed vanga, dark newtonia.

**Reptiles and frogs** Look for Madagascar tree boa, Boettger's chameleon, panther chameleon, two stump-tailed chameleons, *Brookesia tuberculata* and *B. stumpffi*, two leaf-tailed geckos, *Uroplatus alluaudi* and *U. ebenaui*, and two day geckos, *Phelsuma madagascariensis grandis* and *P. lineata dorsivittata*.

## Visitor Information
**Location** 27km south of Diego Suarez (Antsiranana).

**Access** The road is tarred as far as Ambohitra (Joffreville); the last 7km is well-maintained dirt road. In wet months tracks in the park are unsuitable for most vehicles. Footpaths are clearly marked.

**Best months to visit** Accessible and rewarding at any time of year. More comfortable in the drier months (May to November) but better for wildlife between September and April although rain can be heavy.

**Accommodation** Hotels in Diego Suarez; campsite in the park.

**Grading** Easy. Much can be seen in the vicinity of the campsite/car park. Longer trails may be steep and rugged and leeches are a problem after rain.

**Recommendations** Stay at least one night. On a brief visit make the short walk to the Petite Cascade (small waterfall) for a cross section of wildlife.

*Above* Grande Cascade, Montagne d'Ambre National Park. The walk to this waterfall gives a good cross-section of the park's attractions, both plant and animal. *Left* The brilliantly coloured Malagasy kingfisher *(Corythornis vintsioides)* is often seen by the stream leading from the Petite Cascade.

# ANALAMAZAOTRA SPECIAL RESERVE (PÉRINET/ANDASIBE)

Analamazaotra Special Reserve is popularly known by the old French name of the nearby town and railway station, Périnet (Andasibe in Malagasy). Most visitors making the two-day excursion from Antananarivo have one purpose: to hear and see the indri calling, but the reserve is also exceptionally rewarding for reptiles and frogs, and for birdwatching.

The nearby unprotected forests of **Maromizaha** are also gaining popularity. Hiking is strenuous but the rewards in terms of wildlife and views are excellent. The new National Park of **Mantady**, 25km to the north, offers the best chance of seeing the region's rarer species. It is a rugged 2–3 day excursion, and a guide is essential.

## Habitat and Terrain

Périnet, an example of montane rainforest, covers an area of 810ha at altitudes between 930 and 1,040m. Many of the largest trees have been removed and the canopy averages 25–30m. The main part of the reserve resembles a table, with steep, forested sides (up which steps have been cut) bordered by a small lake, and a water course.

The forests at Maromizaha are less disturbed, but have been subject to logging. Nonetheless, there are numerous huge buttress-rooted trees, festooned with mosses and epiphytes.

## Key Species

**Mammals** Indri, brown lemur, grey bamboo lemur, eastern woolly lemur. At night: brown mouse lemur, greater dwarf lemur (summer only). Also red-forest rat and streaked tenrec. If lucky, diademed sifaka and black and white ruffed lemur (Maromizaha and Mantady National Park only).

**Birds** Blue coua, red-fronted coua, blue vanga, red-tailed vanga, velvet asity, sunbird asity, pitta-like ground roller, short-legged ground roller (Moromizaha), brown mesite, collared nightjar, cuckoo roller, Tylas, cryptic warbler.

**Reptiles and frogs** Very good for chameleons, including Parson's chameleon, nose-horned chameleon *(Calumma brevicornis)*, short-horned chameleon *(Calumma nasutus)*, short-nosed chameleon *(Calumma gastrotaenia)*. Also the stump-tailed chameleon *(Brookesia superciliaris)*, leaf-tailed gecko *(Uroplatus fimbriatus)*, and Madagascar tree boa. Frogs include the golden mantella which is found in areas close to the reserve.

## Visitor Information

**Location** 30km east of Moramanga, and approximately 145km east of Antananarivo, off the main road (RN2) between Antananarivo and Tamatave (Toamasina). It is the half-way station on the railway between the capital and main port.

Maromizaha lies to the south of RN2, some 8km from Périnet/Andasibe.

***Access*** From Antananarivo, 4 hours by car, 6 by train. A network of good paths runs throughout the main reserve. Those at Maromizaha are more difficult. The guides here are the best-organised in Madagascar.

***Best months to visit*** September to end of January, April and May. Avoid the cyclone months of February and March. The wildlife is less active during the winter period of June to August.

***Accommodation*** Three simple hotels nearby and several in Moramanga. Camping is permitted outside the reserve.

***Grading*** Easy/moderate. Accommodation is simple and trails steep.

***Recommendations*** An evening and the following morning is sufficient to see the indri. A full day allows you to see other animals. A second day is necessary to explore Maromizaha fully.

*Above* Lac Vert at Périnet (Analamazaotra Special Reserve)

*Left* Indri *(Indri indri)* the 'piebald teddy-bear' whose song enthrals visitors. Indri live in small family groups; their morning calls are answered by groups up to 3km away.

# RANOMAFANA NATIONAL PARK

This is a very beautiful national park, established in the early 1990s to protect the newly discovered golden bamboo lemur. Its pleasant climate, waterfalls and rushing river, and the variety of species make it a deserved favourite.

## Habitat and Terrain

Ranomafana's protected montane rainforest covers an area of 41,600ha, at altitudes between 800m and 1,200m. The area is dominated by the Namorona river which, fed by many streams flowing from the hills, plunges from the eastern escarpment close to the park entrance. The steep slopes are covered with a mixture of primary and secondary forest; much of the secondary growth is dominated by dense stands of introduced Chinese guava and clumps of giant bamboo.

## Key Species

**Mammals** One of the most important mammal sites in Madagascar. Most notable are golden bamboo lemur, greater bamboo lemur, grey bamboo lemur, Milne-Edwards' sifaka, red-bellied lemur, red-fronted brown lemur, and brown mouse lemur. Others include the fanaloka, the eastern ring-tailed mongoose, and the red forest rat. Present but hard to see are fosa and the aquatic tenrec.

**Birds** Blue coua, red-fronted coua, Pollen's vanga, Tylas, velvet asity, sunbird asity, pitta-like ground roller, scaly ground roller, short-legged ground roller, rufous-headed ground roller, brown mesite, Henst's goshawk, Madagascar flufftail, slender-billed flufftail, forest rock-thrush collared nightjar, common jery, green jery.

**Reptiles and frogs** Parson's chameleon, and the similar-looking O'Shaughnessy's chameleon, short-horned chameleon (*Calumma brevicornis*), Madagascar tree boa, three leaf-tailed geckos, *Uroplatus fimbriatus*, *U. sikorae* and *U. phantasticus*, the day gecko *Phelsuma quadriocellata* (common) and numerous frogs belonging to the genera *Boophis* and *Mantidactylus*.

## Visitor Information

**Location** Approximately 65km north-east of Fianarantsoa.

**Access** Via the poor road that connects Fianarantsoa with Mananjary on the east coast, a 2-hour journey from Fianarantsoa. There is an extensive system of trials and paths, but many follow steep and often muddy slopes. The local guides are generally excellent.

**Best months to visit** The summer rainy season (December to March) is the most rewarding, but access and conditions can be difficult. Otherwise the periods either side of the main rains: April and September to November.

**Accommodation** Two rustic hotels in Ranomafana, with another under construction. Camping at various sites within the park.

***Grading*** Moderate. Accommodation is basic and some trails are steep.

***Recommendations*** The minimum stay is one night, allowing an afternoon or night walk and a morning excursion. For enthusiasts, two or three days, with an all-day excursion and one night-time visit. If possible camp for a night in the park to see mouse lemurs and fanaloka, also the scavenging ring-tailed mongoose which visits the campsites.

*Above* Namorona Falls, Ranomafana National Park

*Left* Milne-Edwards' sifaka *(Propithecus diadema edwardsi)*. The diademed sifakas (also known by their Malagasy name, simpona) have ruby-red eyes and are the largest of the sifakas, some subspecies being only slightly smaller than the indri. These animals are being studied in the park, hence the identifying collar.

# THE MASOALA PENINSULA

The peninsula has the largest remaining area of coastal/lowland rainforest on Madagascar, often extending right down to the shore. It lost its protected status in 1964, but plans are well advanced to reclassify around 210,200ha of primary forest lying from sea level to over 1,000m as a new national park. This is currently WWF Madagascar's major conservation priority. From the wildlife point of view it is arguably the most rewarding region in Madagascar.

## Habitat and Terrain

The height of the canopy is around 30m. The slopes are often steep, sometimes almost vertical, and there are numerous clear, fast flowing streams and small rivers. This is the wettest place in Madagascar: the annual rainfall exceeds 3,500mm (and over 5,000mm has been known) and there is no distinct dry season.

## Key Species

*Mammals* This is the only place to see red-ruffed lemur; also present are white-fronted brown lemur, aye-aye, fosa, fanaloka, falanouc, brown-tailed mongoose, greater hedgehog tenrec and lowland streaked tenrec.

*Birds* The most sought after are the helmet vanga, Madagascar serpent eagle and Madagascar red owl. Also brown mesite, red-breasted coua, brown emutail, velvet asity, scaly ground roller, short-legged ground roller, pitta-like ground roller, and Bernier's vanga.

*Reptiles and frogs* Various chameleons including panther chameleon, hooded chameleon *(Calumma cucullata)*, several species of stump-tailed chameleon, tomato frog *(Dyscophus antongili)*, green-backed mantella *(Mantella laevigata)*, two leaf-tailed geckos, *Uroplatus fimbriatus* and *U. lineatus*, and numerous day gecko *(Phelsuma)* species.

## Visitor Information

*Location* The peninsula lies to the east of Maroantsetra and forms the eastern and northern coastline of the Bay of Antongil.

*Access* Only accessible by boat, around 3 hours from Maroantsetra. There are footpaths into the forest around the village of Ambanizana.

*Best months to visit* September to December. November is the most likely month for a few dry days. Masoala is wet throughout the year, but dry spells are possible any time. Avoid the cyclone season (January to March); the boat journey is dangerous in heavy seas.

*Accommodation* Campsites only. All equipment and provisions must be brought from Maroantsetra.

*Grading* Difficult. Hot, often wet, steep trails.

*Recommendations* A visit of a least three days is recommended for this wonderful area. Hotels in Maroantsetra can arrange excursions.

*Above* The Masoala Peninsula. Its future status as national park will protect this beautiful example of lowland rainforest and the numerous endemic species found there.

*Left Heterixalus madagascariensis.* One of the many rare endemic frogs found at Masoala.

# NOSY MANGABE SPECIAL RESERVE

An island in the Bay of Antongil, Nosy Mangabe has been a research centre since 1966 when a group of aye-aye were released here. Many of the species found in Masoala can be more readily seen here and access is easier.

## Habitat and Terrain

Lowland rainforest covers the entire island, some 520ha. It has largely regenerated after being heavily logged around 200 years ago. There are large buttress-rooted trees reaching 35m or more in height. The island's slopes rise steeply from the sea to the summit at 331m.

## Key Species

*Mammals* Aye-aye: with Mananara this is the best place to see these animals. Also black and white ruffed lemur and white-fronted brown lemur.

*Birds* Not particularly diverse or abundant. Watch out for Madagascar pratincole and dimorphic egrets.

*Reptiles and frogs* The best place to see the leaf-tailed gecko *(Uroplatus fimbriatus)*, which is common. Also *Brookesia peyrierasi*, panther chameleon, several day gecko *(Phelsuma)* species, plated lizards *(Zonosaurus madagascariensis* and *Z. aeneus)*. Frogs include the green-backed mantella *(Mantella laevigata)* and other ground-dwelling frogs.

## Visitor Information

*Location* The island lies 5km from Maroantsetra.

*Access* A 45-minute boat ride from Maroansetra. A network of footpaths runs through the forest.

*Best months to visit* Accessible all year around, but avoid the cyclone season when boat journeys are dangerous.

*Accommodation* Well-maintained campsite.

*Grading* Moderate. The steep trails are slippery in wet weather.

*Recommendations* Day trips from Maroantsetra are possible and rewarding, but it is better to spend at least one night to see the nocturnal aye-aye. On day visits the climb to the lighthouse is worthwhile; good for wildlife and a lovely view from the top.

# MANANARA

South of Maroantsetra, opposite the mouth of the Bay of Antongil, is the small town of Mananara. The area was made famous by Gerald Durrell in *The Aye-aye and I*. Verezanantsoro National Park lies within the Mananara-Nord Biosphere Reserve, but it is difficult to get to; visits can be arranged through one of the hotels in Mananara. Closer to the town, in the Mananara river, is 'Aye-aye Island' where sightings of the animals are certain. Visits here must be arranged through the island's owner who runs the hotel, Chez Roger.

Nosy Mangabe

Aye-aye

Mananara

HB

MB

BR

Giant baobabs *(Adansonia grandidieri)* near Morondava

# DRY DECIDUOUS FOREST

The deciduous forests of western Madagascar are less rich in species than the rainforests of the east, but nonetheless contain a wealth of fauna and flora that is of great importance, including some of Madagascar's most endangered animals. The rate of endemicity is higher here than in the east, though the number of species is smaller. The large trees have adapted to a prolonged dry season by shedding their leaves to prevent moisture loss through evaporation. Some, like the baobabs which are such a feature of this area, store water in their bulbous trunks, hence the name 'bottle trees'. Others have roots that are swollen with water to tide them through the drought.

Forests are described as having 'storeys' or layers of vegetation showing different characteristics. Dry western forests have an under storey comprised of very dense shrubs and saplings, many of which keep their leaves during the rainless months, and an upper storey of large trees up to 20m tall which lose their leaves in the dry season. The middle storey has features from both.

Deciduous forest is found on the coastal plain and associated limestone plateaux from sea level to 800m, stretching from Diego Suarez (Antsiranana) at the island's northern tip to Morombe in the southwest. Like the eastern rainforest they have been ravaged by man and are now only found in discontinuous patches, being replaced by largely sterile coarse savannah grassland. This destruction is of particular concern since the trees grow extremely slowly in this zone. The dry season extends from May through to October, but there are considerable differences in rainfall between areas: only about 500mm falls annually in the southwest, whereas the extreme north receives up to 2,000mm.

Within the region there are a number of limestone plateaux that have been eroded into spectacular pinnacle formations known as karst or locally as 'tsingy' (the sound the rock makes when struck). Through these plateaux flow rivers which have created underground passages and caves, some of which have collapsed to form canyons in which forest often flourishes. The main examples of this are at Ankarana and Bemaraha.

Choosing the best time to visit the western forests is difficult. By far the most pleasant months physically are May to September when it is dry and relatively cool, but wildlife is difficult to see. As the rainy season progresses so the humidity rises and insects, particularly flies, become increasingly persistent. When the discomfort is as its worst, December to March, the wildlife – especially reptiles – is at its best. Late September to December is – overall – the most rewarding period.

# ANKARANA SPECIAL RESERVE

This is one of the most exciting reserves in Madagascar – and one of the hardest to visit. Each year more visitors make the effort, however, for the dramatic landscape of limestone pinnacles (tsingy), caves, untouched forest and a wealth of wildlife. Ankarana is thought to contain the highest density of primates of any forest in the world.

## Habitat and Terrain

The Ankarana Massif is a limestone plateau that measures approximately 5km x 20km and rises abruptly from the surrounding grassy plain. It is dominated by impressive formations of 'tsingy'. There is an extensive system of caves (with resident bats), with underground rivers. Some of the largest caves have collapsed, permitting isolated pockets of river-fed forest. Dry deciduous forest grows around the periphery of the massif and penetrates up into the larger canyons.

## Key Species

**Mammals** Crowned lemur, Sanford's brown lemur, northern sportive lemur, fat-tailed dwarf lemur, Amber Mountain fork-marked lemur and western woolly lemur. Also northern ring-tailed mongoose, fanaloka, fosa and several species of bat. Very rare but present are Perrier's black sifaka, western grey bamboo lemur and aye-aye.

**Birds** White-breasted mesite, crested coua, hook-billed vanga, collared nightjar, white-throated rail, Madagascar crested ibis, Madagascar pygmy kingfisher, banded kestrel, Madagascar harrier hawk (Madagascar gymnogene), Madagascar scops owl and pygmy forest kingfisher.

**Reptiles** Oustalet's chameleon, white-lipped chameleon *(Furcifer petteri)*, two big-headed geckos, *Paroedura rarus* and *P. bastardi*, two leaf-tailed geckos, *Uroplatus fimbriatus* and *U. ebenaui*, the day gecko *Phelsuma madagascariensis grandis*, Madagascar tree boa, giant hog-nosed snake and Nile crocodiles in the underground rivers.

## Visitor Information

**Location** Approximately 110km south of Diego Suarez (Antsiranana).

**Access** Reached from Diego Suarez (north) or Ambanja (south). RN6 is paved, and for visitors able to hike the 11km trail (4 hours) to the campsite, access at any time of year is no problem. Camping equipment must be carried – local porters are available. A guide is essential. Groups need a 4x4 vehicle for the 30km track (impassable after rain) to the campsite. There are a few trails in the reserve. Surface water is scarce.

**Best months to visit** Wildlife is more visible in the hot, wet months. Vehicular access is limited to May to October and sometimes the intermediate months of November and April. November is perhaps the most rewarding month.

**Accommodation** Camping only. Surface water is scarce within the reserve.

**Grading** Difficult.

**Recommendations** Stay at least two, preferably three nights. Plan to climb (with a guide) to the Green Lake (Lac Vert) and its surrounding area of 'tsingy'. Tiring, but well worth it.

*Above* 'Tsingy' at Ankarana

*Left* Female crowned lemur *(Eulemur coronatus)* and baby. Crowned lemurs are semi-tame at Ankarana and hang around the campsite looking for handouts (they should not be fed). Ring-tailed mongooses are equally bold.

# AMPIJOROA FORESTRY STATION (ANKARAFANTSIKA STRICT RESERVE)

Ampijoroa competes with Kirindy in providing the best accessible example of western deciduous forest. Birdwatching is outstanding here, access is easy, and a clear system of level paths makes wildlife viewing less of a challenge.

## Habitat and Terrain

The station is part of the Strict Nature Reserve of Ankarafantsika, and is dominated by Lac Ravelobe which lies on the northern side of the road. Around the lake, on sandy soils, grows typical dry deciduous forest. The under storey is sparse, with few epiphytes but abundant lianas. In the rocky, more open areas succulents like elephant's foot plants (*Pachypodium* spp.) and *Aloe* species grow. The terrain is gently undulating or flat, with the occasional shallow ridge.

## Key Species

**Mammals** Coquerel's sifaka, mongoose lemur, brown lemur, western woolly lemur, Milne-Edwards' sportive lemur, fat-tailed dwarf lemur, grey mouse lemur.

**Birds** Best place to see the rare Madagascar fish eagle, and the white-breasted mesite. Also red-capped coua, Coquerel's coua, crested coua, Madagascar green pigeon, Schlegel's asity, Van Dam's vanga, rufous vanga, sickle-billed vanga, greater vasa parrot and Madagascar pygmy kingfisher.

**Reptiles** Oustalet's chameleon, rhinoceros chameleon, stump-tailed chameleon *(Brookesia decaryi)*, iguanid lizards *(Oplurus* sp.*)*, two leaf-tailed geckos, *Uroplatus henkeli* and *U. guentheri*, fish-scaled gecko, Madagascar ground boa, giant hog-nosed snake, spear-nosed snake and the 'fandrefiala' *(Ithycyphus miniatus)*. The lake is home to Nile crocodiles. There is a captive-breeding programme for the plowshare tortoise (angonoka) and flat-tailed tortoise (kapidolo).

## Visitor Information

**Location** The Ampijoroa Forestry Station lies either side of RN4 approximately 120km southeast of Majunga (Mahajunga).

**Access** The road from Majunga is good and the journey takes around 2 hours. Excellent guides are usually available in the reserve.

**Best months to visit** The reptiles are more prominent after the first rains in November. During the summer (December to March), however, it can be extremely hot.

**Accommodation** At present only camping, but simple accommodation may soon be available in the nearby village.

**Grading** Moderate. Easy if taken as a day trip in the cool months.

**Recommendations** Spend at least one night for the best birdwatching and wildlife viewing, or aim to leave Majunga before dawn.

NG

AA

*Above* Lac Ravelobe, Ampijoroa. There is a good chance to see the highly endangered Madagascar fish eagle near this lake.

*Left* Coquerel's sifaka, *(Propithecus verreauxi coquereli)*, and infant. Born in June and July, the baby is at first carried on its mother's front, clinging to her fur. When it is about a month old it rides on her back. These beautiful lemurs are easily seen in Ampijoroa.

# KIRINDY FOREST
# (THE CFPF OR 'SWISS' FOREST)

Kirindy is one of the outstanding wildlife habitats in Madagascar yet it is not a reserve: it is managed by a Swiss company for commercial logging. The CFPF have put their resources into careful research on how to remove economically valuable trees without disturbing the wildlife, and then replant saplings of the same species. In recent years visitor facilities have improved and Kirindy is arguably the best place to see the wildlife of the western region. Several species, such as the giant jumping rat, are only found here.

## Habitat and Terrain

Deciduous forest covering 10,000ha and growing on the sandy soils of the flat western coastal plain at altitudes between 5m and 100m. In addition to the trees found in other deciduous forests, there are three species of baobab, *Adansonia fony*, *A. za* and *A. grandidieri*.

## Key Species

**Mammals** Giant jumping rat, narrow-striped mongoose, fosa. Also notable for six species of nocturnal lemur: pygmy mouse lemur, grey mouse lemur, red-tailed sportive lemur, pale fork-marked lemur, Coquerel's dwarf lemur and fat-tailed dwarf lemur. Diurnal lemurs: Verreaux's sifaka and red-fronted brown lemur. Several tenrec species including common tenrec, lesser hedgehog tenrec and large-eared tenrec.

**Birds** White-breasted mesite, Coquerel's coua, crested coua, sickle-billed vanga, white-headed vanga, rufous vanga, blue vanga, Chabert's vanga, cuckoo roller, grey-headed lovebird, Madagascar nightjar, Madagascar harrier hawk, Henst's goshawk and banded kestrel.

**Reptiles** Flat-tailed tortoise or kapidolo, Dumeril's boa, Madagascar ground boa, giant hog-nosed snake, spear-nosed snake, plated lizards (*Zonosaurus* sp.), Oustalet's chameleon, one-horned chameleon (*Furcifer labordi*), two leaf-tailed geckos (*Uroplatus fimbriatus* and *U. guntheri*), big-headed gecko and several day gecko (*Phelsuma*) species.

## Visitor Information

**Location** Approximately 50km northeast of Morondava.

**Access** The drive from Morondava, via 'the Avenue of Baobabs' takes about 1½ hours on a good road. Within the park are wide, level trails.

**Best months to visit** October to December, before it gets too hot.

**Accommodation** A well-maintained campsite only but visitor bungalows are planned.

**Grading** Moderate. Easy walking, but very hot.

**Recommendations** One or two nights are needed to see the nocturnal animals which are Kirindy's speciality.

*Above* A broad trail in Kirindy

*Left* The giant jumping rat *(Hypogeomys antimena)* locally known as 'Vositse', is only found in this small area of Madagascar. When in a hurry it jumps like a small wallaby, but when searching for its favourite food of roots and saplings it walks normally on all fours.

Didiereacaea

# THE SOUTHERN REGION

This area is perhaps the most peculiar and unusual in Madagascar. It extends from Morombe in the southwest right around the southern coast to Fort Dauphin (Tolagnaro) and inland for a distance up to 50km. The vegetation comprises a type of deciduous thicket or thorn scrub dominated by members of the Didiereaceae and Euphorbiaceae families. This is commonly referred to as 'spiny forest' or 'spiny desert'.

Didiereaceae resemble cacti, although they are only very distant relatives. This is an excellent example of parallel evolution where nature comes up with similar water-retaining, predator-deterring solutions to dry climates, even though their plant families are different.

Mixed in with the four genera of Didiereaceae are Euphorbias. Most ooze latex if cut or damaged, and only a few species bear thorns.

Conspicuous among these strange plants are the 'bottle trees'. These include baobabs, which tower above the thickets, and species of *Pachypodium* or 'elephant's foot' – some really do look like an elephant's foot, being squat and grey, whilst others look remarkably like a spiny bottle. All in all it's a most fascinating landscape.

Because of its arid nature, this area has suffered less than others on the island from 'slash and burn' agriculture. Nonetheless, the forests are becoming increasingly fragmented as the population increases with intensified pressures on the forests for charcoal and building materials. While some protected areas have been set aside, there is an urgent need for a reserve to be created in the southwestern spiny forest to the north of Tuléar. Some highly localised and threatened birds, reptiles and plants are found here.

Dry gallery forest, also called riverine forest or tamarind forest, occurs by southern rivers. This is superficially similar in appearance to western dry forest, but is dominated by huge tamarind trees *(Tamarindus indica)*, locally called 'kily', which may exceed 20m. There are also sprawling banyan *(Ficus)* trees as well as an under storey of shrubs and saplings. The famous private reserve of Berenty, home of the ring-tailed lemurs, is mainly gallery forest.

Ring-tailed lemur

31

# BERENTY PRIVATE RESERVE

This is the best-known reserve in Madagascar. For most visitors it is infinitely rewarding, with a combination of comfortable accommodation, friendly lemurs, and easy walking in the gallery forest.

## Habitat and Terrain
The reserve is situated on the banks of the Mandrare river and covers an area of 265ha. Adjacent to the gallery forest are areas of spiny forest. The terrain is flat.

## Key Species
*Mammals* Ring-tailed lemur, Verreaux's sifaka, red-fronted brown lemur (introduced), collared brown lemur (introduced), grey mouse lemur, white-footed sportive lemur, greater hedgehog tenrec, Madagascar fruit bat.

*Birds* Giant coua, hook-billed vanga, Madagascar magpie-robin, Madagascar paradise flycatcher, ashy cuckoo shrike, souimanga sunbird, grey-headed lovebird, two species of vasa parrot, Frances's sparrowhawk, Madagascar harrier hawk, Malagasy scops owl, white-browed owl.

*Reptiles* Radiated tortoise, spider tortoise, Madagascar ground boa, Oustelet's chameleon, warty chameleon, jewel chameleon, big-headed gecko, plated lizard *(Zonosaurus trilineatus)*, the near-limbless lizard, *Androngo trivittatus*.

## Visitor Information
*Location* About 80km west of Fort Dauphin (Tolagnaro), just north of Amboasary.

*Access* With sightseeing stops, the journey from Fort Dauphin takes 2 hours on a good road. There is an excellent network of wide paths and trails throughout the reserve. Guides are available, but not mandatory.

*Best months to visit* Accessible all year round. September and October for baby lemurs. During and after the rains (December to March) for reptiles.

*Accommodation* A very pleasant complex of bungalows with a good restaurant. This is a private reserve run by the de Heaulme family: it is generally necessary to stay at one of their hotels in Fort Dauphin (Tolagnaro) and arrange a visit from there.

*Grading* Easy.

*Recommendations* An expensive reserve. Feasible as a day trip, but much better to stay over night. Best wildlife viewing is at dawn. Also take a night walk to look for invertebrates as well as nocturnal lemurs.

A cheaper alternative to Berenty is **Amboasary-Sud**, another private reserve operated by the Kaleta group of hotels. Again, it is necessary to stay at one of these hotels to arrange a visit. There is a campsite but no other facilities. This reserve is more suited to the independent traveller or as a day trip from Fort Dauphin. It is located to the south of Berenty, and is similar in terms of habitat and species seen.

*Above* A troop of inquisitive ring-tailed lemurs observe the photographer on a typical Berenty trail. Ring-tails in Berenty are completely fearless and will climb all over visitors.

*Left* Giant coua *(Coua gigas)*. Berenty provides rewarding birdwatching with several of Madagascar's endemic species easily seen.

# AROUND TULÉAR (TOLIARA): IFATY AND ANAKAO

Many visitors to Tuléar stay in the string of beach hotels at Ifaty, north of the noisy port. Just inland is a good example of spiny forest, and an excursion there is worthwhile, especially for birdwatchers.

South of Tuléar is Anakao, another popular beach area.

## Habitat and Terrain

Ifaty is typical spiny forest, dominated by species of Didiereaceae and the baobab species local to the region, *Adansonia rubrostipa*. The whole area is very flat and sandy. Waders and other sea birds are attracted to the mud-flats, rocky areas and beaches around the coast. At Anakao the vegetation is thorn scrub/thicket with *Euphorbia* species very prominent.

## Key Species

*Mammals* With luck: ring-tailed lemur, grey mouse lemur, white-footed sportive lemur, lesser hedgehog tenrec, Madagascar fruit bat (Anakao).

*Birds* A number of rare local endemics: long-tailed ground roller, sub-desert mesite, Lafresnaye's vanga, Archbold's newtonia, thamnornis warbler, sub-desert brush warbler, running coua, Verreaux's coua, banded kestrel, Madagascar plover, littoral rock-thrush.

*Reptiles* Radiated tortoise, spider tortoise, Dumeril's boa, spiny-tailed iguanid lizards *(Oplurus* spp.*)*, three-eyed lizard *(Chalaradon madagascariensis)*, near limbless lizards *(Pygomeles braconnieri* and *Androngo trivittatus)*, the bark gecko, *Homopholis sakalava*, horned chameleon *(Furcifer antimena)*.

## Visitor Information

*Location* The beach complex of Ifaty is about 30km north of Tuléar (Toliara). Anakao is a fishing village about 60km south of Tuléar, accessible by boat or ferry (or by very poor road).

*Access* Ifaty is reached by dirt road from Tuléar (over an hour). The boat trip to Anakao generally takes 1½ hours. There are paths through the forest, but they are not marked and are often overgrown so it is easy to get lost.

*Best months to visit* October to January. In the hot, wet season (December to March) the roads can be very poor. The cool season is more pleasant but fewer birds and reptiles are seen.

*Accommodation* Both places have good hotels. At Ifaty there are several beachfront hotels with good restaurants. At Anakao there is a single hotel which is more basic.

*Grading* Easy/moderate. The intense heat is a problem for some people.

*Recommendations* Good guides are available from the hotels. They are generally well-informed on birds, reptiles and botany. Visits without a guide may seem uninteresting and are potentially dangerous. Ifaty walks should be limited to the hours around dawn and dusk; at other times it is too hot and you will see little. Go to the coral reefs instead.

*Above* 'Spiny forest' near Ifaty. In the foreground is the baobab of the region, *Adansonia rubrostipa*, behind which is a species of Didiereaceae, *Alluaudia ascendens*.

*Left* Long-tailed ground roller *(Uratelornis chimera)*. This rare endemic bird inhabits only a small area of the south-western spiny forest.

# ISALO NATIONAL PARK

Isalo is quite unlike any other place in Madagascar. Its appeal is the remarkable landscape of eroded 'ruiniforme' sandstone formations, rare plants and the feeling of space and stillness.

## Habitat and Terrain

The park covers 81,540ha of the Isalo Massif which rises up from the flat surrounding grassy plain. The sandstone has been eroded into strange shapes, cut through by impressive gorges and canyons. Vegetation is concentrated in the canyon bottoms where streams flow. These wooded areas are dominated by the fire-resistant tree, tapia *(Uapaca bojeri)*, on which a Malagasy endemic silkworm feeds, along with *Pandanus pulcher* and the locally endemic feather palm, *Chrysalidocarpus isaloensis*. On the cliffs and rocks are several endemic succulents including the elephant's foot and the Isalo aloe *(Aloe isaloensis)*.

## Key Species

*Mammals* Not prominent, but these may be seen: ring-tailed lemur, Verreaux's sifaka, red-fronted brown lemur.

*Birds* Benson's rock-thrush, white-throated rail, Madagascar coucal, Madagascar wagtail, Madagascar kestrel.

*Reptiles and frogs* Oustalet's chameleon, jewel chameleon, spiny-tailed iguanid lizard *(Oplurus saxicola)*, a stump-tailed chameleon, *Brookesia ebenaui*, and two locally endemic frogs, the beautifully coloured *Scaphiophryne gottlebei* and the brownish *Mantidactylus corvus*.

## Visitor Information

*Location* Between Fianarantsoa and Tuléar (Toliara) on RN7, to the north of the village of Ranohira.

*Access* Reached from Ihosy, approx 90km (2 hours) to the east, or from Tuléar (Toliara) 250km to the southwest (3 to 4 hours). A local guide is obligatory if hiking within the park.

*Best months to visit* Brief rains occur between January and March. Daytime temperatures from June to August are pleasant but nights can be very cold; from November to March days may be too hot. During September and early October the elephant's foot plants are in bloom and temperatures are moderate.

*Accommodation* Camping only within the park. Two basic hotels in Ranohira and a more comfortable one at the southern edge of the massif.

*Grading* Easy (day trips); moderate to difficult if camping and hiking.

*Recommendations* Even driving through Isalo gives some idea of its rugged beauty. To see any wildlife, however, you must hike and camp. A popular area is the natural swimming pool, *La Piscine Naturelle*, and for a better chance of lemur-viewing the Canyon des Singes and an extension to the beautiful Grotte de Portugais are worth the effort.

NG

GT

*Above Pachypodium rosalatum* var. *gracilis* in Isalo National Park

*Left* Benson's rock-thrush *(Pseudocossyphus bensoni)*. This rare bird is endemic to Isalo and surrounding area. Its plumage allows it to blend in with the lichen-covered sandstone, but males can often be seen singing in the mornings from prominent rock perches.

# ZOMBITSE FOREST

This pocket of forest on the road to Tuléar is mainly for a birdwatching stop – there are some rare and very localised species here. Other wildlife such as lemurs may also be seen, which makes a visit all the more worthwhile if you are not able to go to the other southern reserves.

## Habitat and Terrain

Zombitse (and its neighbour to the northeast, Vohibasia) constitute the last remnants of transition forest between the western and southern floristic domains. Two baobabs, *Adansonia madagascariensis* and *A. za*, are conspicuous, but the canopy generally averages only 15m. Zombitse covers an area of 21,500ha but the forest margins have been eaten into by agriculture. The soil is sandy and the terrain flat.

## Key Species

*Mammals* Hard to find because of poaching. Verreaux's sifaka, ring-tailed lemur, red-fronted brown lemur, grey mouse lemur, red-tailed sportive lemur, pale fork-marked lemur, fat-tailed dwarf lemur, fosa, common tenrec, large-eared tenrec, greater and lesser hedgehog tenrec.

*Birds* Appert's greenbul, long-billed greenbul, long-billed green sunbird, thamnornis warbler, white-headed vanga, sickle-billed vanga, blue vanga, Chabert's vanga, giant coua, crested coua, Madagascar lesser cuckoo, Madagascar hoopoe, cuckoo roller, Madagascar sandgrouse and Madagascar partridge.

*Reptiles* Madagascar ground boa, Dumeril's boa, Oustalet's chameleon, warty chameleon, big-headed gecko *(Paroedura sp.)*.

## Visitor Information

*Location* Straddling RN7, some 25km east of Sakaraha.

*Access* From Tuléar (Toliara) 2 to 3 hours by car, from Ranohira 1 hour. No formal paths, though plans to develop Zombitse as a national park include the provision of basic facilities and the training of guides.

*Best months to visit* October to April.

*Accommodation* There are no nearby hotels and campers will need to be entirely self sufficient (including water supplies).

*Grading* Moderate.

*Recommendations* Most people make short stops en route to the coast. For the enthusiast, it is worth camping for one or two nights.

*Left* Appert's greenbul *(Phyllastrephus apperti)*. This species is found only in Zombitse and Vohibasia forests and is one of five endemic greenbuls.

PM

# MAMMALS

Verreaux's sifaka
*(Propithecus verreauxi verreauxi)*

# LEMURS

Lemurs, like ourselves, are primates. Lemurs, however, are regarded as 'primitive' since they share characteristics with early ancestral primates. For this reason they are known as prosimians, or pre-monkeys; other prosimians include the nocturnal bushbabies, lorises and tarsiers of Africa and Asia. In contrast to these, many lemur species are active during the day (and periodically during the night) and live in family groups or large troops in which the females are dominant. This is rare in primates: with most monkeys and great apes the males are larger and indisputably the boss.

Another major difference between the lemurs and their more intelligent monkey relatives is an acute sense of smell, and some species have long, dog-like noses. Scents and smells are important in lemur society and are used extensively for communication, information gathering and marking territories.

By the time the first primates evolved, some 60 million years ago, Madagascar had already broken away from Africa, so the Mozambique Channel would have been a formidable challenge for any animal to cross, especially if, like lemurs, it disliked water. Perhaps the early lemurs floated on rafts of vegetation, perhaps they crossed via a series of now submerged islands, or perhaps – and this is a recent theory – they hopped on to Madagascar as the island was reunited with mother Africa during a northward drift before heading for its current position in the Indian Ocean.

We will probably never know the answer. What we do know is that lemur competitors and most predators missed the boat, which gave the early prosimian colonists a free hand to diversify and exploit every available niche of this huge island. Today no less than 50 different varieties survive (and at least 15 other species are known to have become extinct after the arrival of man). They range in size from the tiny pygmy mouse lemur, which could sit in an eggcup, to the indri, a piebald teddy bear weighing around 7kg.

This great diversity (five families and 14 genera) fascinates biologists who marvel that an island this size can hold over a third of the world's primate families. To other visitors the appeal is more fundamental; lemurs are among the most cuddly, endearing and bewitching animals in existence. Their soft fur, round bright eyes and gentle black-gloved hands give lemurs an irresistible appeal.

NG

### Lemurs great and small

*Left* The pygmy mouse lemur *(Microcebus myoxinus)* weighs as little as 25g, and is probably the smallest primate in the world. First named in 1852, it 'disappeared' for over a century through confusion with other species, to be rediscovered and identified in the early 1990s.

*Below* The indri *(Indri indri)* is disputably the largest lemur and is certainly the most vocal. Indri are easily seen in Périnet where their haunting song provides one of Madagascar's unforgettable experiences.

NG

# THE INDRI FAMILY: INDRIIDAE

The indri family comprises the indri, sifakas, and woolly lemurs. *Indri indri* live in small family groups in the northeast rainforests. Indri territory is too big to defend by scent alone, so their song proclaims their whereabouts and warns others to keep away. The sifakas (genus *Propithecus*) are the most widely distributed and diverse members of the family and are the favourite with visitors.

Indri and sifakas have somewhat human proportions, with long, powerful legs and shorter arms. They are superb leapers, jumping effortlessly from tree to tree, but are awkward on the ground. Whereas visitors seldom see indri descend from their trees, sifakas frequently need to cross open ground. They bound on their hind legs, bellies thrust out and arms aloft (see opposite), providing observers with one of Madagascar's comic spectacles.

Of the sifaka species, Verreaux's sifaka is the most common. It is divided into four subspecies, two of which are easily seen. *Propithecus verreauxi verreauxi* (page 39, below and opposite) is one of the main attractions of Berenty, where it feeds on leaves, buds, fruit and flowers. It appears never to drink, and is one of the few mammals at home in the arid Didiereaceae forest where it can leap on to the spiny boughs without damaging its hands or feet.

*Below* Verreaux's sifaka *(Propithecus verreauxi verreauxi)* sometimes supplement their diet in unusual ways. This female with her infant (the young are born in August and September) is eating a large fungus on the forest floor at Berenty.

NG

## Other Sifakas

Coquerel's sifaka *(Propithecus verreauxi coquereli)* is the other *P. verreauxi* subspecies that is easy to see. Its chestnut-coloured arms and thighs make it particularly handsome. These sifakas are found in the dry forests of the northwest, and in Ampijoroa a family group often lounges in the trees close to the entrance to the reserve. The young are born in June and July.

The other two subspecies, the crowned sifaka *(Propithecus verreauxi coronatus)* and Decken's sifaka *(Propithecus verreauxi deckeni)*, are confined to more remote areas of western forests and are difficult to see.

Madagascar's eastern forests are the domain of the simpona, *Propithecus diadema*, also divided into four subspecies which are quite dissimilar in appearance. The diademed sifaka *(Propithecus diadema diadema)* is considered by many to be the most beautiful of all the lemurs: its silky coat is a combination of orange, gold, white and black, and it has piercing ruby red eyes. It is also the largest sifaka and perhaps the largest lemur, sometimes exceeding the weight of the indri. Unfortunately it is an elusive animal, but those with the energy for a long hike into Mantady National Park or Maromizaha, near Périnet, have some chance of seeing it.

Two other elusive subspecies are *Propithecus diadema candidus*, which is pure white, and *Propithecus diadema perrieri*, which is jet black. Easiest to see, however, is *Propithecus diadema edwardsi*, Milne-Edwards' sifaka (see page 17).

The third sifaka species is the golden crowned sifaka *(Propithecus tattersalli)* which is confined to a tiny enclave of forest in the far north (see *Discoveries*, page 54).

## Woolly Lemurs

The two species of woolly lemur, *Avahi laniger* and *Avahi occidentalis*, are the only nocturnal members of the family, although they are often seen during the day sleeping in trees or shrubs. They may be mistaken for sportive lemurs *(Lepilemur spp.)* until the distinctive white patches on the back of the thighs are noticed. Woolly lemurs are low-energy animals living on a low-energy diet of – mainly – leaves. Even at night they spend much of their time resting.

*Right* Eastern woolly lemur *(Avahi laniger)* in its daytime resting place at Périnet.

CH

Coquerel's sifaka *(Propithecus verreauxi coquereli)*

NG

Diademed sifaka *(Propithecus diadema diadema)*

NG

# THE 'TRUE LEMURS': FAMILY LEMURIDAE

These are perhaps the most familiar lemurs, being frequently seen in zoos. The most captivating is surely the ring-tailed lemur (*Lemur catta*), instantly recognisable and synonymous with Madagascar. Ring-tails are the most terrestrial of all lemurs, as visitors to Berenty soon realise when troop members scamper across the sand and surround arriving vehicles. Ring-tails live in troops of around 20 animals in a female-dominant society where scent plays an essential part. Females can be observed rubbing their anal glands against the base of trees; males do the same but also use a spur and gland on their wrists to gouge the bark and enforce the troop's 'keep out' signs. Males also indulge in 'stink fights': after anointing their tails with scent from their wrist glands, they stand glaring at their opponents, their tails quivering aloft like smelly black and white flags.

Ring-tails have a variety of calls, some used to warn of danger (with different calls for aerial and terrestrial dangers. The morning call is like the mewing of a cat (hence the name *catta*). Their varied diet ranges from fruit, leaves and flowers to insects and the occasional reptile. Unlike the sifakas which share their range, they also need to drink so prefer gallery (riverside) forest.

Ring-tailed lemurs are found throughout the southwest of Madagascar. Although Berenty is the easiest place to see them, the neighbouring reserve of Amboasary-Sud is equally rewarding. Isalo and Beza-Mahafaly are other possibilities.

*Below* Ruffed lemurs. The two subspecies of *Varecia variegata* are uncommon in their eastern rainforest habitat where they are sometimes hunted for food. Black and white ruffed lemurs (*Varecia variegata variegata*) are found over quite a large area of north and central eastern rainforest, but the red ruffed lemur (*Varecia variegata rubra*) is confined to the Masoala Peninsula. The Antainambalana River forms a natural barrier between the two subspecies, but in captivity hybrids between the two are quite common.

Twins and sometimes triplets are born in September and October. Baby ruffed lemurs do not cling to their mother but are parked in nests until they are old enough to follow her around the forest canopy.

*Varecia variegata variegata*

*Varecia variegata rubra*

*Above* Ring-tailed lemur (*Lemur catta*).
These lemurs generally give birth in
August and September, sometimes to
twins. Initially the babies cling to the fur
of the mother's belly, later riding on her
back. The white youngster *(left)* was born
at Berenty in 1995. It has been named
Sapphire by researchers because of its
blue eyes, but it is not a true albino since
it has retained the distinctive ringed tail.

## Genus *Eulemur*

Widespread throughout the forests of Madagascar, with the exception of spiny forest, the *Eulemur* species share several characteristics. Most noticeably, almost all are sexually dichromatic: males are a different colour from females. Scent also plays an important part in marking territories; females generally spray urine, while males smear secretions from their anal region or heads on to strategic objects – which may include the females. Although more at home in trees, most species spend some time on the ground, strutting about on all fours with their bottoms in the air.

Five species of *Eulemur* are recognised, although some of these are divided into several subspecies. The species used as the benchmark to describe the genus is the rather rare mongoose lemur *(Eulemur mongoz)* which can sometimes be seen at Ampijoroa. Its behaviour pattern is unusual: during the rainy season it is active during the day but with the onset of the dry season (around June) it switches to being nocturnal until the December rains.

## Classification of 'True Lemurs'

Lemur taxonomy (the science of arranging living things into hierarchies of related groups: species, genera, families, etc) has undergone a series of recent revisions. At one time the genus *Lemur* contained all the species now decribed as *Lemur, Eulemur* and *Varecia*. However, most authorities now regard the ring-tailed lemur *(Lemur catta)* and ruffed lemurs *(Varecia)* as sufficiently different from the others to warrant separation, hence the recent creation of the genus *Eulemur*.

*Right* Red-bellied lemur, *Eulemur rubriventer*. The white patches in front of the eyes show it to be a male. Females have less white on their faces but pale underparts. Both sexes have a dark or black tail. Unusually, these lemurs live in monogamous pairs and both males and females carry the infants on their backs. Ranomafana is the easiest place to see them.

NG

*Above* A male crowned lemur *(Eulemur coronatus)*. The male's orange-brown coat and black crown are distinctive and contrast with the female (see page 25) which is grey with an orange tiara. Crowned lemurs are only found in the far north of Madagascar, and may be seen in Montagne d'Ambre National Park and Ankarana.

*Left* Female black lemur *(Eulemur macaco macaco)* in Lokobe Reserve, on the island of Nosy Be. The difference in colour between the sexes is most pronounced in this sub-species: males are all black with bright orange eyes, females are chestnut brown with white ear tufts. The rare Sclater's black lemur *(E. m. flavifrons)*, which lives in a tiny area of the northwest, has blue eyes and no eartufts.

JA

## Brown Lemurs: *Eulemur fulvus*

The brown lemurs are divided into no fewer than six subspecies. In all of these the females are boringly alike, generally with a brown body and black or grey head, whilst the males show off with a variety of markings, colours and facial hair. These handsome animals not only look different from their mates but from the males of other subspecies. Each subspecies lives in its own distinct range and these join together to form a ring around Madagascar's periphery.

The distribution of the common brown lemur *(E. f. fulvus)* is certainly interesting. It is found in both the central eastern rainforests and the dry deciduous forests of the northwest. When the forest cover on Madagascar was more extensive, it seems likely this lemur was found in the adjoining areas across the central highlands. There is also an isolated population on the island of Mayotte in the Comores. This group was almost certainly introduced by man, and once thought to be yet another subspecies, '*E. f. mayottensis*', but this is no longer considered to be so.

The males of two subspecies in particular are notable for their splendid white whiskers. The white-fronted brown lemur *(E. f. albifrons)*, shown opposite is the most handsome and is found in the forests of the northeast. Almost as impressive is the male Sanford's brown lemur *(E. f. sanfordi)* which has long cream-coloured eartufts and side-whiskers. The females of both subspecies are a similar uniform brown. However, although both occur in northern Madagascar, their ranges do not overlap so there is little chance of confusion. Sanford's brown lemur is best seen at Montagne d'Ambre National Park and Ankarana, where it shares the forests with the crowned lemur *(Eulemur coronatus)*.

*Right* The common brown lemur *(Eulemur fulvus fulvus)*. Although hardly 'common' it is readily seen at Ampijoroa and Périnet. Unlike other subspecies, the two sexes are very similar in appearance.

White-fronted brown lemur
*(Eulemur fulvus albifrons)*. The male,
with his snowy-white head and
creamy underparts, is gorgeous;
the female *(left)* is less impressive.
Nosy Mangabe is the best place to
see them in the wild, but an easy
option is to visit the free-range
troop at Ivoloina Zoo, near
Tamatave (Toamasina).

JB

QB

*Above right* Red-fronted brown lemur *(Eulemur fulvus rufus)*. This is the brown lemur most commonly seen by visitors - it occurs in the southeast (Ranomafana) and the southwest (Kirindy) but the best opportunity for a close view is at Berenty where it has been introduced. Males can be recognised by their greyish coat with a rufous crown, whilst females (pictured here) are pale chestnut brown with a grey crown. Both sexes have conspicuous white patches above their eyes and long dark noses.

*Above left* Collared lemur *(Eulemur fulvus collaris)*. This subspecies has also been introduced to Berenty where it has been much less successful than *E. f. rufus* although the two occasionally mate: there are a few hybrids.

Brown lemurs generally live in troops of between five and 15 animals with males and females in roughly equal numbers. Most prefer fruit as the mainstay of their diet, along with leaves and flowers. However, at Ranomanfana National Park, the red-fronted brown lemur *(E. f. rufus)* has been observed eating giant millipedes, first wiping off any unpleasant secretions with its tail, before tucking into the unlikely snack.

Although all brown lemurs are generally considered to be diurnal, many of them (and other *Eulemur* species) are known to be up and about at night as well. The extent of these nocturnal antics may vary with the seasons and is probably also influenced by the cycle of the moon – around full moon they tend to be more active. Animals that are active both day and night are said to be cathemeral.

Cathemeral : active both by day and night

# BAMBOO OR GENTLE LEMURS

The genus *Hapalemur* is known by various popular names of which 'gentle lemur' is least appropriate for an animal that does not hesitate to bite. However, these lemurs are most endearing animals with thick, greyish fur and blunt faces. As their other common name implies, they are associated with bamboo thickets, and in some areas have become quite common in secondary forests where bamboo flourishes.

The grey bamboo lemur *(Hapalemur griseus)* occurs as three subspecies. The eastern grey bamboo lemur *(H. g. griseus)* is found throughout the rainforest belt and is often seen at Périnet and Ranomafana. Around Lac Alaotra in the northeast it gives way to the Lac Alaotra bamboo lemur *(H. g. alaotrensis)*, which is unique amongst primates in that it lives and feeds on reeds. However, it is also critically endangered as the reedbeds around the lake are destroyed to make way for agriculture. The third subspecies, *Hapalemur griseus occidentalis*, is, as its name suggests, confined to remote areas of western forest where it is unlikely to be seen.

*Above* The eastern grey bamboo lemur *(Hapalemur griseus griseus)* is the most widespread of the bamboo lemurs.

*Left* The Lac Alaotra bamboo lemur *(Hapalemur griseus alaotrensis)* is perhaps the most restricted of the *Hapalemur* genus. This very rare lemur can also be seen at Jersey Zoo and Duke University Primate Centre (USA).

*Left* Golden bamboo lemur *(Hapalemur aureus)*. The coat is a beautiful rich brown, lightening to an orange-golden chest, neck, cheeks and eyebrows. Relatively new to science, this species is one of Madagascar's most endangered lemurs. Its diet consists almost entirely of giant bamboo which is laced with sufficient cyanide to kill most mammals, including humans. Such selective feeding may help it avoid competition with the other two bamboo lemurs that share its range.

Visitors willing to spend several days trudging the trails at Ranomafana National Park could hit what might be described as the 'bamboo lemur jackpot': this is the only place where all three species are known to occur together. Spotting a grey bamboo lemur is quite likely, but finding the other two takes a great deal of luck and expert guidance. The prize is a glimpse of the recently discovered golden bamboo lemur *(Hapalemur aureus)* and the equally endangered greater bamboo lemur *(Hapalemur simus)*.

## NEW DISCOVERIES

No fewer than four species of lemur have recently been discovered or rediscovered. The golden bamboo lemur was first spotted in Ranomafana in 1986 and identified as a new species in 1987. It is probably also found further south, at Andringitra. The golden-crowned sifaka *(Propithecus tattersalli)* was seen by scientists in 1982 in northern Madagascar, but not described as a new species until 1988. Since then no other populations have been found and it remains one of the most threatened lemurs. It is confined to a precariously small range around Daraina and enjoys no protection at present. Similarly the tiny hairy-eared dwarf lemur *(Allocebus trichotis)* was known from only four museum specimens until its rediscovery in 1989.

And most recently, in 1992, the pygmy mouse lemur *(Microcebus myoxinus)* resurfaced. It was formally identified as a separate species in 1994 (see page 41).

Hairy-eared dwarf lemur *(Allocebus trichotis)*

# Nocturnal Lemurs

*Grey mouse lemur (Microcebus murinus)*

## THE MOUSE LEMURS AND DWARF LEMURS: FAMILY CHEIROGALEIDAE

**Mouse Lemurs (genus *Microcebus*)**
Madagascar's tiny, and very lively, mouse lemurs are the most abundant of the island's primates, perhaps outnumbering humans. The grey mouse lemur, *Microcebus murinus* (page 55), lives in the drier forests of the south and west, while the brown mouse lemur, *Microcebus rufus* (page 124), prefers the wetter east. The pygmy mouse lemur, *Microcebus myoxinus* (page 41), has so far only been observed in Kirindy and the nearby reserve of Analabe.

**Dwarf Lemurs (genus *Cheirogaleus*)**
There are two species: the greater dwarf lemur *(Cheirogaleus major)*, which lives in the east, and the fat-tailed dwarf lemur *(Cheirogaleus medius)*, which is found in the west and south. The easier one to see is *C. major* which, in the warm season, is common along the roadside near Périnet. During the cold winter months dwarf lemurs hibernate (aestivate), living off the fat in their tails, a unique feature among primates. Even when active they are slow moving, particularly in contrast with the scampering mouse lemurs.

**Other Family Members (genera *Mirza, Phaner* and *Allocebus*)**
The remaining members of the family are represented by Coquerel's dwarf lemur *(Mirza coquereli)*, the fork-marked lemur *(Phaner furcifer)* and the hairy-eared dwarf lemur *(Allocebus trichotis* – see page 54). Coquerel's dwarf lemur lives in the dry western forests, and behaves like a mouse lemur but is much larger. The four subspecies of fork-marked lemur are distributed in the forests of both the east and west. They can be identified by the black line along the back, which forks at the head, ending at each eye.

## WEASEL OR SPORTIVE LEMURS: FAMILY MEGALADAPIDAE

The generic name, *Lepilemur*, is more appropriate for general use than the choice of official English names: lepilemurs don't look like weasels and are not very sportive. They are most often seen during the day, peering dozily out of their sleeping holes in tree trunks (see front cover). At night, however, they are often very vocal and active.

Despite their small size, lepilemurs are related to some of the large extinct lemurs from the genus *Megaladapis*, although they differ enough to warrant their own subfamily, Lepilemurinae.

Seven species are known, their ranges forming an almost continuous ring around Madagascar. Many are very similar in appearance so geographical location is usually the most reliable means of identification.

*Lepilemur leucopus*, which is familiar to visitors to Berenty, is also found in

Caecotrophy : The behavioural habit of animals that eat their own faeces to allow their digestive tract a second chance to absorb the few nutrients available.

QB

QB

BR

GT

*Top* Fat-tailed dwarf lemur *(Cheirogaleus medius)*, Ampijoroa.

*Above left* Grey-backed sportive lemur *(Lepilemur dorsalis)*, Lokobe. There are few tree holes in this reserve so these sportive lemurs are usually seen during the day resting in the fork of a tree.

*Above right* White-footed sportive lemur *(Lepilemur leucopus)*, Berenty.

*Left* Milne-Edwards' sportive lemur *(Lepilemur edwardsi)*, Ampijoroa.

*Front cover* Red-tailed sportive lemur *(Lepilemur ruficaudatus)*, Kirindy.

# AYE-AYE: FAMILY DAUBENTONIIDAE

There is only one member of this family, the aye-aye *(Daubentonia madagascariensis)*. This is a seriously weird animal which was only classified as a lemur in relatively recent times: originally scientists took it to be a squirrel-like rodent. What makes it so unusual? It has a disproportionately long and bushy tail; it has teeth like a rodent – they never stop growing; its ears are like those of a bat – but even larger and more mobile – and so sensitive that it can apparently hear a grub moving under the bark of a tree; it has claws not fingernails (except on the great toe); it has inguinal teats (between its back legs not on its chest); it has no fixed mating season but can give birth at any time of the year. And it has that extraordinary finger.

Most descriptions of the aye-aye say that the middle finger is 'greatly elongated'. In fact it is no longer than the middle finger of other primates, including man. But it *is* extraordinarily thin, skeletal in fact, and this makes it look longer. The aye-aye also keeps the other fingers crooked up out of the way when working with its most important digit, so its hand looks like a tarantula spider. The thin finger is designed to fit through the gnawed holes in tree branches or large nuts (coconuts are now a favourite) and winkle out the tasty contents, while all the time the aye-aye's ears move to pick up the slightest sound.

If you can't see an aye-aye in the wild, try to see one in a zoo (Tsimbazaza in Antananarivo, or in Jersey, Paris, or Duke University Primate Center, USA). Remember it is nocturnal so even a good sighting in the wild will give you a less than perfect view. However, aye-ayes seem to be popping up all over Madagascar, and the chances of seeing one are improving. Mananara and Nosy Mangabe (see page 21) are still the best places, but one has even been seen by the road at Périnet.

*Left* The aye-aye using its thin middle finger to extract grubs from a tree cavity.

Aye-aye *(Daubentonia madagascariensis)*

MP

# VIVERRIDS: THE CIVETS AND MONGOOSES

## CARNIVORE CONFUSION

There are only eight species of carnivore in Madagascar (a low number compared with other mammal groups); all are endemic, and belong to the family Viverridae, more commonly known as civets, genets and mongooses. These animals share certain basic characteristics, which suggests a common ancestor – probably of African origin – though evolution has since taken them down very different paths.

There are three subfamilies of civet-like carnivores, each containing a single species: Cryptoproctinae – the fosa *(Cryptoprocta ferox)*, Fossinae – the fanaloka or Malagasy striped civet *(Fossa fossana)* and Euplerinae – the falanouc *(Eupleres goudotii)*.

The common names for these animals are very confusing. *Fossa fossana*, the Malagasy striped civet, is often confused with the fosa *(Cryptoprocta ferox)*. Presumably, this mistake was first made by Gray in 1864 when he gave the civet its generic name, *Fossa*. To add to the muddle the striped civet's Malagasy name, fanaloka, is pronounced 'fanalook' and sometimes even written 'fanalouc' with inevitable confusion with falanouc. The last straw is that the Malagasy themselves use the name fossa or fosa (pronounced foosa or foosh) as a general term which can refer to a number of the island's carnivores! We, at least, will be consistent: fosa, fanaloka, falanouc.

*Left* A female falanouc *(Eupleres goudotii)* with her young.

*Above* Fosa *(Cryptoprocta ferox)*. The fosa is Madagascar's largest carnivore, measuring about 2m in length, half of which is tail. It is a secretive, mainly nocturnal animal, rarely seen by visitors. This individual was photographed at Kirindy, where the animals are being studied.

*Below* Fanaloka or Malagasy striped civet *(Fossa fossana)*. This animal is much smaller than the fosa, shy and strictly nocturnal. It is attracted to the smell of cooking and often seen near the campsite at Ranomafana and also at Ankarana.

# CIVET-LIKE CARNIVORES

The fosa resembles an elongated, short-legged puma (when first discovered it was thought to be a member of the cat family), although its general build is much more slender: it rarely weighs more than 10kg. Active mainly between dusk and dawn, fosas are extremely agile climbers, perfectly at home bounding around the larger canopy branches. Lemurs make up more than half their diet.

The fanaloka or Malagasy striped civet is a small, spotted, fox-like carnivore the size of a domestic cat. Fanalokas live in pairs in the eastern rainforests, foraging for food in the dense undergrowth. In preparation for the leaner winter months, they are able to lay down fat reserves, especially in the tail.

Madagascar's most specialised carnivore is the falanouc. This uncommon and secretive animal lives in the lowland rainforests of the east and northeast. It is larger than the fanaloka and has an extended snout and tiny teeth – features that help it catch the earthworms and other invertebrates which are its exclusive diet. The claws and forepaws are well developed for digging and are also used for defence. The falanouc gives birth to extremely well-developed young: babies are born with their eyes open and are able to follow their mother and hide in vegetation within two days.

# MONGOOSES

Madagascar's mongooses, of which there are five species, all belong the endemic sub-family Galidiinae. The commonest and most widespread species is the ring-tailed mongoose *(Galidia elegans)*, which is split into three subspecies which inhabit different forest regions. All have a rich russet coat and distinctive banded tail. They are sociable creatures, often found in vocal family groups. Equally at home on the ground or in the branches of trees, they forage for rodents, young birds, eggs, reptiles and invertebrates.

The narrow-striped mongoose *(Mungotictus decemlineata)* is found only in the dry areas of the west. These delightful grey/sandy coloured animals have several faint dark stripes along their flanks and back and a large bushy tail which is held erect when alarmed.

The little-studied brown-tailed mongoose *(Salanoia concolor)* lives in the northeastern rainforests. It is known to be active mainly during the day and to feed mostly on insects.

The broad-striped mongoose *(Galidictis fasciata)*, of the eastern rainforests, is the most specialised flesh-eater in the subfamily. It is mainly nocturnal, feeding on rodents, lizards and frogs. Its cousin, the giant striped or Grandidier's mongoose *(Galidictis grandidieri)* was only discovered in 1986, after the examination of mislabelled museum specimens. It is the largest and least known mongoose species, and may be restricted to the spiny forest areas around Lac Tsimanampetsotsa in the southwest.

*Above* Northern ring-tailed mongoose *(Galidia elegans dambrensis)*. These pretty animals often visit the campsites in reserves such as Ankarana.

*Below* Narrow-striped mongoose *(Mungotictus decemlineata)*. Locally called 'boky-boky', this mongoose is limited to the dry western forests, particularly Kirindy.

# TENRECS

Tenrecs (family Tenrecidae) are insectivores, a group of mammals which have flourished in Madagascar, branching into at least 26 species. Tenrecs are divided into two sub-families: Tenrecinae, the so-called spiny tenrecs (five species), and Oryzorictinae, the furred tenrecs, with about 21 species.

Not all the 'spiny' tenrecs live up to their name. The largest, the tail-less or common tenrec *(Tenrec ecaudatus)*, has only a few spines hidden in its fur. It weighs up to 2kg and is a popular source of meat for the Malagasy (see page 100). The reproductive capabilities of the common tenrec are remarkable: it can produce up to 31 embryos at one time and has 24 nipples. Not all the babies survive, but those that do follow their mother around the forest in orderly columns, wearing stripy, spiny coats for camouflage and defence.

The greater hedgehog tenrec *(Setifer setosus)* and lesser hedgehog tenrec *(Echinops telfairi)* look very much like hedgehogs, and also roll themselves into a prickly ball. The former is found in both western and eastern forests, while the latter is restricted to the drier regions of the southwest.

In the rainforest areas, the most commonly encountered species are the streaked tenrecs: *Hemicentetes semispinosus* from the lowlands, and *H. nigriceps* from the highlands. Both have sharp yellow spines mixed in with softer black prickles arranged in longitudinal stripes, and orange-yellow underparts. A specialised set of dorsal spines is used for communication: they can be vibrated together to produce a threatening rattle (called stridulation) or an inaudible (to humans) sound used to call straying youngsters or family members.

The 'furred' tenrecs have evolved to fill niches occupied elsewhere by shrews, moles or desmans. There are probably more than 16 species of shrew tenrec (genus *Microgale*), which can sometimes only be distinguished from each other by close examination of their teeth! By far the most interesting and remarkable member of this subfamily is the very rare aquatic tenrec *(Limnogale mergulus)*, which has developed webbed feet and a flattened tail for swimming. It lives in fast-flowing streams, where it forages for frogs, fish, crustaceans and aquatic insect larvae.

Aquatic tenrec *(Limnogale mergulus)*

*Above left* Lowland streaked tenrec *(Hemicentetes semispinosus)*. These little yellow and black striped tenrecs are often seen foraging for earthworms in eastern forests or by the roadside. The formidable cream-coloured spines are used for head-butting attackers.

*Above right* Large-eared tenrec *(Geogale aurita)*. This mouse-like inhabitant of the western forests uses its keen hearing to locate termites in rotting trees. Like other tenrecs it may aestivate (go into a torpor) during the dry months when food is scarce.

*Below* Lesser hedgehog tenrec *(Echinops telfairi)*. Surprisingly, for such a clumsy-looking animal, this tenrec spends much of its time in trees in the western forests and is an agile climber. Like other tenrecs of the region, it aestivates during the dry season.

# RODENTS

Madagascar has some charming rodents, all of which belong to the endemic subfamily Nesomyinae. Only 19 have so far been classified. The most celebrated Malagasy rodent is the giant jumping rat *(Hypogeomys antimena)* which is restricted to a small area in the west (see page 29). This charming animal is the size of a rabbit and fills the same ecological niche: it lives in family groups and digs a nework of burrows.

Perhaps the easiest species to see is the red forest rat *(Nesomys rufus)*, which may be active during the day and is common in many rainforest areas. It is slightly smaller than the ubiquitous brown rat, and its diet consists mainly of fallen fruits and seeds. In typically adaptable rodent fashion, it is also regularly seen foraging around rubbish dumps at campsites and seems to have little fear of humans.

Two other large rainforest rats belonging to the genus *Brachytarsomys* are unusual in being totally tree-dwelling, with prehensile tails for gripping branches. They are nocturnal. There are also eight species of smaller tree-living rats from the genus *Eliurus*. Their tails are not prehensile, but often end in a conspicuous brush-like tuft, hence their alternative name, tuft-tailed rat.

The smallest of the island's rodents is the mouse, *Macrotarsomys bastardi*, which is similar in many ways to the gerbils of north Africa, having large ears, long hind feet and a long tail with a bristly tip.

*Below left* The inquisitive red forest rat *(Nesomys rufus)* is often seen around campsites.

*Below right Macrotarsomys bastardi*, a gerbil-like mouse from the western forests.

# BATS

ats (order Chiroptera) are split into two very distinct groups: the fruit bats, or flying foxes (Megachiroptera), and the insectivorous-type bats (Microchiroptera). Both these groups are represented in Madagascar, although they are the least studied of the island's mammals. Given the relative ease with which bats can travel over long distances (and more importantly over water), it is hardly surprising that there are fewer endemic species of bats here than in the other groups of mammals.

There are three species of fruit bat, of which the Madagascar fruit bat *(Pteropus rufus)* is the largest – its wing span may reach 1.5m and it can weigh over 1kg. The name flying fox is appropriate – they do indeed look like little foxes, with long muzzles and very mobile ears. They are generally found in large colonies in forests and often on islands around the coast where there is a small patch of forest.

The second largest Malagasy fruit bat is *Eidolon dupreanum*, which prefers to roost in caves and rock crevices and is restricted to the western deciduous forests and some higher montane areas. The Madagascar rousette fruit bat *(Rousettus madagascariensis)* is unusual amongst fruit bats in possessing basic echolocation capabilities, allowing it to navigate in low light and to roost in caves. This species is widely distributed in the eastern rainforest areas.

*Above* The Madagascar fruit bat or flying fox *(Pteropus rufus)* is a familiar sight in some areas. Noisy colonies congregate in favourite roost trees in reserves such as Berenty, and on islands like Nosy Tanikely *(right)*. They are often active during the day, taking off en masse to circle their roost before hanging upside down from a branch. This bat has relatives on other Indian Ocean islands, but otherwise its nearest family members are in Asia – there are no *Pteropus* species in Africa.

Six families from the suborder Microchiroptera are represented in Madagascar (there are 19 worldwide), although only one of these, Myzopodidae, is endemic. It contains a single species, the sucker-footed bat *(Myzopoda aurita)*. This rare bat is found in the eastern rainforest and derives its name from peculiar suction discs on the wrists and feet, which help the animal to hang from the smooth leaves of palms, where it likes to roost.

Although known as 'insectivorous-type' bats, by no means all Microchiroptera eat insects. As more research takes place in Madagascar the number of species will, no doubt, be added to. To date 20 or so have been described, most of which also occur on mainland Africa.

*Below left* Madagascar red trident bat *(Triaenops persicus rufus)*. This endemic sub-species of leaf-nosed bat has three 'prongs' above the 'nose leaf' instead of the more usual one. The prongs and 'leaf' are part of the bat's echolocation system, an extraordinarily complex means of navigation used by many bats which is still imperfectly understood. Possibly the prongs act as a sound splitter, directing the echo of the bat's squeaks – or sounds made by its prey – into stereophonic sound to give better direction finding. The 'leaf' and ears act together to ensure that both prey-detection and navigation can take place simultaneously.

*Below right* Madagascar mouse-eared bat *(Myotis goudoti)*. This bat does not have a leaf-nose, but in front of each ear is a projection or *tragus*, which may have an equivalent function to the nose prong in leaf-nosed bats. This bat is endemic to Madagascar and the unusual russet colour on the upper parts possibly affords some camouflage on the so-called 'red island'.

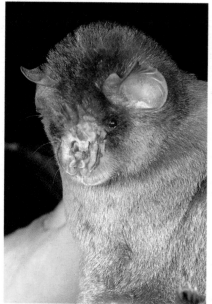 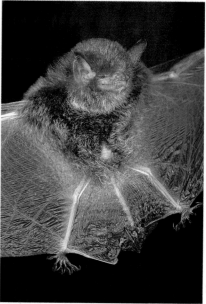

# BIRDS

Pitta-like ground roller *(Aterlornis pittoides)*

# THE BIRDS

## ENDEMIC ODDITIES

Madagascar has only 258 bird species, but is home to 36 endemic genera – more than any other country in the African region. Here it is quality not quantity that attracts birders. There are 110 endemic species, with five endemic families and one endemic sub-family. The vast majority are more or less dependent on the native forests or wetlands and very few inhabit the more open central highlands.

Madagascar shares 25 species with either the Comoros or Aldabra. Many of these belong to genera confined to the western Indian Ocean islands, such as the greater and lesser vasa parrot, Madagascar blue pigeon, Madagascar magpie robin and the fodys. Happily it is the species present on Madagascar which are plentiful, in contrast to their endangered cousins on the other islands.

The beautiful bird on page 69 is the pitta-like ground roller *(Aterlornis pittoides)*. The ground-rollers (Brachypteraciidae) are a most attractive family. There are four rainforest species and one from the spiny forest, the long-tailed ground roller (see page 35). All excavate long nest burrows, and feed on large insects and small reptiles.

Identifying birds can seem a daunting business, so here are two endemic species that are unmistakable.

*Left* The crested drongo *(Dicrurus forficatus)*, is black with a deeply forked tail and comical 'crest'.

*Below* The Madagascar red fody *(Foudia madagascariensis)*, one of four endemic weavers, dresses to impress the ladies, so outside the breeding season is a boring brownish colour. Between November and April, however, its brilliant (mostly) red plumage makes it stand out from the crowd.

*Left* Crested coua *(Coua cristata).* The most widely spread coua, it is found mainly in western dry forests but it is also present in the eastern rainforests.

All nine species of the handsome coua family are easy to see. Couas are related to cuckoos and coucals but are sufficiently distinct to warrant their own subfamily, Couinae. Despite being forest birds, only three magpie-size species live in the trees: crested coua, blue coua, and Verreaux's coua *(Coua verreauxi).* These birds bear a striking resemblance to Africa's touracos. All couas share two conspicuous characteristics: featherless, blue skin around the eyes and long, broad tails. The six ground-dwelling species behave in a similar manner to the American roadrunner or the Old World pheasants. The largest is the rather stately giant coua *(Coua gigas),* shown on page 33, which is often heard before it is seen. The running coua *(Coua cursor)* may be encountered sprinting through the spiny forest. It has an area of black skin on the rump which it exposes to the rays of the early morning sun after a particularly cold night. The red-capped coua *(Coua ruficeps)* and Coquerel's coua *(Coua coquereli)* which inhabit the western and southern areas, also warm themselves in this way. The red-fronted coua *(Coua reynaudii)* and red-breasted coua *(Coua serriana)* are both rainforest birds.

*Above* Blue coua *(Coua caerula).* This bird of the rainforest may also occasionally be seen in Ankarana.

## Primitive Parrots

Unlike their very rare cousin on the Seychelles, the black vasa parrots (*Coracopsis* spp.) of Madagascar are commonly seen. Some experts consider these to be the most primitive of all the family Psittacidae. Certainly they are very different from the more familiar parrots, being some of the least colourful but the most tuneful. Vasa parrots, like other parrots, also have a repertoire of raucous squawks but are mostly located because of their liquid whistles, which may be heard even on moonlit nights.

In the breeding season some greater vasa parrots lose their head feathers and their skin turns orange-yellow.

## The Birds and the Beaks

The vangas (Vangidae) are undoubtedly the most celebrated endemic bird family in Madagascar. Had Charles Darwin sailed to this island instead of the Galapagos, and seen vangas instead of finches, his thoughts on species and evolution would surely have been similarly provoked. Such is the diversity in size, colour and in particular beak shape of the 14 species that it is hard to imagine they are related at all. Yet characteristics of the skull and other structural features confirm a common ancestry similar to the helmet shrikes of Africa. Today they fill niches that are occupied, in other parts of the world, by woodpeckers, woodhoopoes, shrikes, tits, treecreepers and nuthatches – all birds absent from Madagascar.

The size and shape of the various vanga beaks reflect the size of their insect prey, its location and the mode of capture. Larger species, like the hook-billed vanga, have a robust beak with a characteristic hook at the tip, to help them deal with their carnivorous diet of large insects, chameleons and other small vertebrates. The extraordinary helmet vanga comes into this category. Others, in the *Xenopirostris* genus such as Pollen's vanga, use their laterally compressed beaks to rip bark from dead wood. The more lightly built Chabert's vanga behaves like a flycatcher, grabbing insects on the wing.

The sickle-billed vanga is the largest and most easily identifiable species. Its babbling, cackling call, gregariousness and feeding behaviour are similar to Africa's woodhoopoes. It lives in the western forests where its bill is used for probing crevices in the bark of trees.

The smaller species, which include the red-tailed vanga *(Calicalicus madagascariensis)* and nuthatch vanga *(Hypositta corallirostris)*, have much finer beaks suited to hunting out small insects along branches and tree trunks. The 'Tylas vanga' *(Tylas eduardi)* has now been removed from this family, having been reclassified as a flycatcher. It mimics the plumage of the far more robust Pollen's vanga but has a much finer bill. This mimicry probably acts as a deterrent to small raptors which would be reluctant to attack a vanga which could use its powerful beak in defence. Most vangas are gregarious and are often seen in large mixed feeding flocks; they are one of the most successful endemic families on the island.

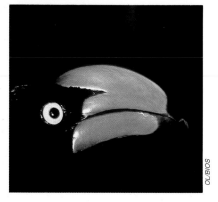

*Above left* Chabert's vanga *(Leptopterus chaberti)*

*Above right* Hook-billed vanga *(Vanga curvirostris)*

*Centre left* Pollen's vanga *(Xenopirostris polleni)*

*Centre right* Sickle-billed vanga *(Falculea palliata)*

*Below left* Blue vanga *(Cyanolanius madagascarinus)*

*Below right* Helmet vanga *(Euryceros prevostii)*

*Above* Brown mesite *(Mesitornis unicolor)*.
The mesites (Mesitornithidae) are a peculiar
family of ground dwelling birds that rarely fly
even when pursued by predators. Instead they
freeze, relying on their cryptic colouring to
help them blend into the background. There
are only three species: the brown mesite,
found in the eastern rainforest, the white-
breasted mesite *(Mesitornis variegata)* a
deciduous forest bird, and the sub-desert
mesite *(Monias benschi)* from the spiny forest.

*Right* The asity family (Philepittidae)
shows striking differences between males and
females. In breeding plumage, the males of all
four species sport iridescent blue-green wattles,
caruncles and naked areas of facial skin.

*Above right* Male sunbird asity *(Neodrepanis
coruscans)* from the smallest of the two genera.
The sunbird asity's close resemblance to true
sunbirds illustrates the principle of convergent
evolution: if a design works, repeat it.

*Below right* Male velvet asity *(Philepitta castanea)*.
These lovely birds, found in the eastern
rainforest, are sometimes seen in Périnet and
(more often) at Ranomafana.

# RARITIES AND REDISCOVERIES

Madagascar red owl *(Tyto soumagnei)*

# THE BRINK OF EXTINCTION?

Until very recently several of Madagascar's endemic birds were considered to be on the brink of extinction. Sadly this is still true of the Delacour's grebe (Aloatra little grebe), *Tachybaptus rufolavatus*, and the world's rarest duck, the Madagascar pochard, *Aythy innonata*, but the recent surge of interest in Madagascar and dramatic increase in research has revealed many pleasant surprises.

Perhaps the two most spectacular examples are those of the Madagascar serpent eagle *(Eutriorchis astur)* and the Madagascar red owl *(Tyto soumagnei)* shown on page 75. The serpent eagle had eluded detection for 50 years, until a dead specimen found in 1990 confirmed its existence. The red owl had not been seen since 1973. A spate of research in the 1990s revealed that both species are alive and well, particularly in primary rainforests of Masoala peninsula where they have been recorded several times and even tracked by means of radio transmitters. The red owl has also been seen in Mantady so is likely found in other large rainforests. Thus, as large as they are, these birds represent two excellent examples of species which were merely overlooked for decades.

The 1995 sighting of the Sakalava rail *(Amaurornis olivieri)* at Lac Bemamba in western Madagascar effectively settled the last avian mystery in Madagascar. This secretive small black crake has suffered from the loss of its wetland habitat. The lake is also the main haunt of another rarity, the Madagascar or Bernier's teal *(Anas bernieri)*, which is the westernmost species of the cluster of Australasian grey teals. The slender-billed flufftail *(Sarothrura watersi)* was another bird thought to be on the threshold of extinction but it is now sighted regularly, particularly around Ranomafana.

Certain other species are threatened as their ranges are very limited. None more so than Appert's greenbul that lives only in the transition forests of Zombitse and Vohibasia (see page 38). Also in danger are the sub-desert mesite and long-tailed ground roller which share a narrow, unprotected range north of Tuléar.

Despite the continued habitat loss new bird species are still being discovered. Since 1990, two new names have been added to the ornithological register. The first, the sub-desert brush warbler *(Nesillas lantzii)*, is found in the spiny forest areas to the north of Tuléar, while the second, the cryptic warbler of the eastern rainforests, is sufficiently distinctive to warrant the creation of its own genus, *Cryptosylicola*.

*Left* Madagascar serpent eagle
*(Eutriorchis astur)*

# REPTILES
# AND FROGS

Boettger's chameleon *(Calumma boettgeri)*

# REPTILES

**M**adagascar is a wonderful place for herpetologists (who study reptiles and amphibians), with around 300 known reptile species and more being added almost every month as new surveys are done. Over 90% are endemic.

Although you would expect Madagascar to share most of its reptile families with Africa, many well-known ones from the mainland are absent. For instance, among lizards there are no agamas, no 'typical' lizards of the Lacertidae family, and no monitors. On land there are no front-fanged venomous snakes (which means, in effect, there are none of the deadly species such as vipers, cobras and mambas) and no pythons. Some families have their closest relatives in South America: the iguanid lizards and the boas.

## CHAMELEONS: FAMILY CHAMAELEONIDAE

Chameleons are mostly found in Africa and Madagascar. About half the world's 135 or so species are unique to Madagascar and new ones are still being discovered.

Chameleons are perhaps the most distinctive and specialised of all lizards, perfectly designed for life in the trees. Evolution has done an exquisite job here, modifying almost every bodily feature (the *Brookesia* genus is the exception – see page 84). A chameleon's body is laterally flattened, enabling the animal to move easily through tangles of branches and allows them to absorb heat efficiently in the morning and evening by turning broadside to the sun. This shape provides an imposing profile to deter predators and other chameleons (they are all strictly solitary). An angry or frightened chameleon puffs itself up to look even bigger.

It is the chameleon's eyes that most people notice first. Large, and protected by circular eyelids which cover all but the pupils, they can be swivelled independently so the reptile can look in two directions at once without needing to move its head. This is extremely useful for an animal that relies on camouflage to avoid predation: it can keep absolutely still but watch out for danger – and food – in all directions. When food is located (usually an insect), both eyes point forward. The chameleon can then judge depth and distance and bring into action its most specialised feature – its tongue. This rests like a primed missile in the chameleon's mouth before being fired at potential prey. The tongue can extend to a length equal to the animal's body, and if the aim is good a sticky muscular tip clamps onto the victim and seals its fate.

## Perfect adaptation to life in the trees

*Above left* Warty chameleon *(Furcifer verrucosus)*. Using its prehensile tail as a fifth hand, the chameleon is an agile climber. Chameleons seem to prefer to rest on thin branches; perhaps they are safer from heavier predators, although exposed to a passing bird – or human.

*Above right* Parson's chameleon *(Calumma parsonii)*. A chameleon's toes and fingers are fused together in two opposing groups like a pair of pliers – an ideal adaptation for gripping branches. The independently moving eyes can be seen here: one eye keeps the photographer in view and the other watches where the chameleon is going.

*Below* Panther chameleon *(Furcifer pardalis)*. The speed with which the tongue shoots out to hit its target has been measured at less that a quarter of a second.

## Goliaths and midgets

The world's largest and smallest chameleons. Two species compete for the heavyweight title: Oustalet's chameleon *(Furcifer oustaleti)*, *top*, from the drier areas of the west and southwest, and Parson's chameleon *(Calumma parsonii)*, *centre*, which lives in the eastern rainforests, where four forms are recognised. Both these giants can exceed 60cm in length (including the tail). Oustalet's chameleon varies a lot in colour; many are a dull shade of grey-brown.

The nose-horned chameleon *(Calumma nasutus)*, *below*, is the smallest of the 'typical' chameleons. Its total length rarely exceeds 100mm, and much of that is tail. However, the most minuscule chameleon of all is a pygmy stump-tailed chameleon *(Brookesia minima)*, *centre*, which is barely longer than a fingernail, 35mm in total length.

## A Coat of Many Colours

Chameleons do not change colour to match their background! It is just one of the several beliefs that have been attached to this extraordinary lizard through the centuries. In ancient times they were probably kept as pets in southern Europe: Aristotle described their ability to change colour and Shakespeare and his contemporaries claimed they fed on air (the caged chameleons presumably caught insects that came within tongue-shot).

In the chameleon's world, colour is a language used to defend territories, convey emotions and communicate with potential mates. It is also a means of regulating body temperature. The way the change in colour is achieved is fascinating. Cells containing a variety of pigments lie underneath the skin and and are able to 'open' and 'close' to expose their pigment. Colour change is controlled by a combination of hormonal and nervous activity. For instance, a distressed or angry chameleon opens cells containing the brown pigment, melanin, which turns it much darker. When the chameleon relaxes, yellow cells and blue cells combine, resulting in the calmer more normal shades of green. Sexual excitement produces an explosion of colours and patterns. At night many chameleons turn almost white, perhaps the result of total relaxation.

*Above* A panther chameleon crossing the road gives lie to the belief that chameleons change colour to match their background. This male is in the breeding colours that he adopts during the wet season when mating and egg-laying take place. Females, which, in this species, are a pinkish brown colour, need soft, moist soil to dig holes for their eggs. Chameleons are awkward on the ground and adopt a swaying motion which perhaps makes them less conspicuous to a normal predator, but enables human admirers to catch them with ease.

*Right* A defensive female one-horned chameleon *(Furcifer labordi)* in Kirindy. This chameleon is putting on the works: she is already in her breeding colours, which have intensified under stress. In addition, she inflates herself with air, dilates her throat, and opens her mouth to show its bright orange interior.

## A Nose for All Occasions

Anything Cyrano or Pinocchio could do, chameleons can do better. The array of weird and wonderful noses displayed by Madagascar's chameleons is fantastic, from small rounded bumps to twin prongs and long slender lances. Although often called 'horns', they are in fact scale-covered extensions of the nose, and are known scientifically as rostral protuberances. Generally only the males have them: they are used, like colour, in combat and to impress the females.

Noses are useful for others of the same species, especially when related species live in close proximity to one another. Experiments have shown that, if the protuberances are removed, the individual chameleon can be thrown into total confusion and is not recognised by its own kind.

*Below* A courting Will's chameleon *(Furcifer willsii)* shows off his range of seductive colours. The female is dressed in her 'ready and willing' garb of green. After mating, hormonal changes will prompt the female to change livery to a much darker shade. If another male tries to have a go, her cloak of disapproval darkens again and continues to be worn until after her eggs are laid, reinforcing the message that she is no longer available.

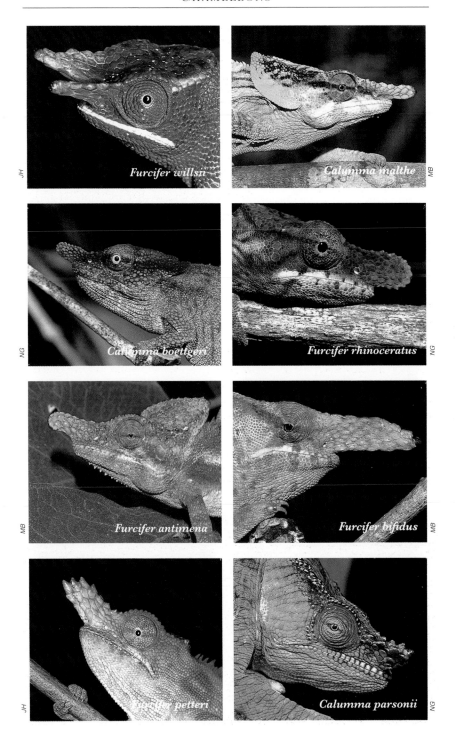

*Furcifer willsii* — JH

*Calumma malthe* — MB

*Calumma boettgeri* — NG

*Furcifer rhinoceratus* — NG

*Furcifer antimena* — MB

*Furcifer bifidus* — MB

*Furcifer petteri* — JH

*Calumma parsonii* — NG

**Chameleons in Miniature**

The stump-tailed or leaf chameleons from the genus *Brookesia* are the most diminutive of all chameleons, ranging in size from 110mm down to just 30mm. They are also almost totally terrestrial, spending the majority of their time on the forest floor. Only at night do some species climb up into the lower twigs of undergrowth to sleep. Their appearance reflects their preferred habitat – they are beautifully camouflaged when among leaves. Although the rainforests are their stronghold, certain *Brookesia* species are also found in some dry deciduous forest areas.

Though so tiny, and superficially so different from 'typical' chameleons, *Brookesia* share all the hallmark chameleon characteristics although some of these are reduced: their ability to change colour is limited and their tail is short and only partially prehensile.

A common trait amongst the stumptail chameleons is to feign death if threatened. Some individuals fold their legs underneath their bellies and roll on to their side to resemble a dead leaf. Others flatten themselves laterally and produce rapid body vibrations to try and deter a would-be attacker. As a group, they are clearly successful: 24 *Brookesia* species are currently known and new ones are still being found.

*Brookesia stumpfii*

*Brookesia decaryi*

MB

*Brookesia perarmata*

MB

*Brookesia superciliaris*

MB

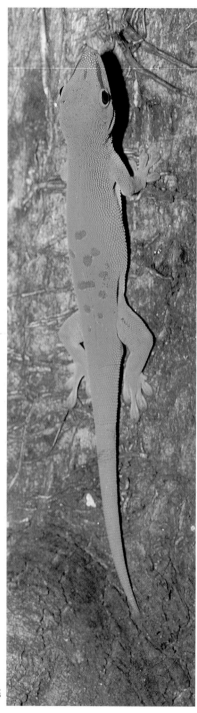

## GECKOS: FAMILY GEKKONIDAE

In terms of species richness, geckos are amongst the most successful of all lizards, with about 800 species known worldwide; in Madagascar geckos outnumber all other lizard groups. The majority of geckos have specialised scales on their feet with microscopic hooks which allow them to cling to vertical surfaces. In most species the large eyes are protected by single transparent scales which are periodically licked clean by the gecko's long, flattened tongue.

They are generally nocturnal and brown in colour, but day geckos (genus *Phelsuma*) have been described as the 'living jewels of Madagascar' and have radiated into numerous flamboyant and beautiful species. The majority are emerald green with a variety of red, orange, blue or dark spots and blotches on their head, back and flanks. The largest is *Phelsuma madagascariensis grandis (left)* which reaches 30cm or more.

The big-headed geckos live on the ground and at night can be heard rustling around in the leaflitter. There are nine species on the island, widely distributed around the various types of forest. Those from the west and south, such as *Paroedura pictus* (shown on the opposite page) and *P. bastardi*, are the most likely to be seen. *P. masobe*, the largest of the genus and said to be the most beautiful, was only discovered in 1994 in low elevation rainforest.

*Above* Fish-scaled gecko *(Geckolepis typica)*. These lizards are easily identified by their very large overlapping scales. Their response to predators is startling. If seized, they can shed their entire coat of scales in an instant, exposing the bare skin underneath, and make good their escape. The flayed gecko is a horrendous sight but the scales soon regenerate.

*Below* Big-headed gecko *(Paroedura pictus)*. These nocturnal geckos can be heard moving around on the forest floor throughout the island.

## The Remarkable Uroplatus

The leaf-tailed or fringed geckos (genus *Uroplatus*) are amongst Madagascar's most extraordinary animals. There may be no other vertebrates in the world that demonstrates such mastery of camouflage. The largest species, *Uroplatus fimbriatus*, (shown on the left and on page 131), perhaps illustrates this best. The skin colour and pattern exactly mimics its favoured tree trunk. The gecko rests motionless, head downward, stretching out its legs and spatula-like tail and flattening itself against the bark. A frill of skin around the lower part of the animal forms a continuous skirt which blends the outline of the gecko imperceptibly into the tree. To enhance the effect, these geckos can also change colour should they find themselves on a lighter or darker tree.

If this camouflage fails, *Uroplatus fimbriatus* has one more remarkable defensive trick up its sleeve. When alarmed it flicks its tail upwards, throws back its head and opens its mouth as wide as possible, showing a brilliant orange-red interior. Another species, *U. henkeli*, adds a startlingly loud distress call to the gaping mouth threat.

At present ten species are recognised, from the largest *U. fimbriatus*, which can measure 30cm, to the smallest *U. ebenaui*, which is about 7–8cm long. They are more abundant in rainforests, but are also found in deciduous forests. The larger species (page 89) tend to mimic tree bark while the smaller ones (page 90) look like dry leaves.

*Uroplatus henkeli*

WL

*Uroplatus alluaudi*

NG

**Uroplatus sikorae**

WL

MB

**Uroplatus lineatus**

WL

**Uroplatus phantasticus**

NG

**Uroplatus ebenaui**

# OTHER LIZARDS

Besides the chameleons and geckos, Madagascar is home to three other lizard families: the iguanids, plated lizards and skinks.

### Iguanids: Family Iguanidae

As has already been mentioned, the presence of iguanids presents something of a riddle as the stronghold of this group of lizards is Central and South America. In Madagascar they mainly inhabit the west and southwest of the island where the climate is hot and dry. The small, three-eyed lizard *(Chalaradon madagascariensis)* is particularly common in the dry south and in the spiny forest. Its 'third eye' is a pineal eye which shows as a conspicuous black dot on the back of its head. This 'eye' is found in many lizards and in some species is sensitive to light, possibly measuring periods of day and night.

The spiny-tailed iguanids, genus *Oplurus*, are common in the dry south and west of Madagascar. Their tails, which look like elongated fir cones, are used as a defensive barrier to their hiding places. These large lizards spend the day on rocks and trees, waiting for an insect meal to walk past. They also eat fruit and leaves.

*Below* Collared iguanid *(Oplurus cuvieri)*. Normally these large lizards seem quite placid and are easy to approach. The skirmish here is probably over territory.

## Plated (Girdle-tailed) Lizards and Skinks:
## Families Gerrhosauridae and Scincidae

The smooth, streamlined lizards belonging to these two families are sometimes mistaken for snakes as they move through the leaf-litter. Some species of skink add to this impression by having only vestigial legs.

Plated lizards are common throughout Madagascar. There are 13 species of *Zonosaurus*, all of which are found on the forest floor where they can be heard scurrying through the leaves.

Skinks are a familiar and large family of lizards (700–800 species worldwide) but are perhaps the least studied of Madagascar's reptiles. Around 50 species are currently recognised. Many skinks have small limbs, and the near-limbless lizard, *Androngo* genus, has almost dispensed with them altogether. These burrowing skinks are sometimes seen at Berenty.

*Above* Plated lizard *(Zonosaurus laticaudatus)*. This lizard was photographed at Ampijoroa, but the species is widespread throughout Madagascar.

*Below* An unidentified *Amphiglossus* species from Ranomafana. Like *Amphiglossus astrolabi*, this large skink is unusual because of its aquatic habits – it is often found in streams.

# SNAKES

*Above* One of 85 species of snake found in Madagascar. This is an undescribed one (genus *Stenophis*) known from only a few specimens, and first collected in the forest of Beroboka in the southwest. There are many snakes in this genus awaiting formal identification.

*Left* This beautiful snake is probably a juvenile *Pseudoxyrhopus microps*.

## The Boas: Family Boidae

Boas are primarily a South American group of snakes, though also found on Madagascar. There are three species, all of which show very close links to the boa constrictor from Central and South America.

The largest is the Madagascar ground boa *(Acrantophis madagascariensis)*, which averages just under 2m in length but can reach 3m. Its geometric pattern of browns, creams, greys and black, helps the snake blend into the leaf-litter. The ground boa hunts mainly at night and is particularly fond of small mammals (including lemurs). In the drier western forests, this species may, like some other snakes, spend the day in the underground burrows of ant colonies. Dumeril's boa *(Acrantophis dumerili)* is a closely related species from the south.

The Madagascar tree boa *(Sanzinia madagascariensis)* is the smallest and most common of the family. There is considerable variation is colour between snakes from different areas: those from the east sport a crazy-paving pattern of olive-green, grey and black, whereas those from western areas are much more brown. The juveniles are also a totally different colour (see below).

*Left* Madagascar tree boa *(Sanzinia madagascariensis)*. Despite its name, it is often found on the ground. Much darker specimens are sometimes seen, with an almost iridescent blue sheen. A row of heat sensitive pits around their upper and lower lips help it to find warm-blooded prey.

*Below* A juvenile of the same species. Its brilliant red colour will change to green as it matures.

*Above* The spear-nosed snake *(Langaha madagascariensis)* is an excessively weird sight in an island full of such surprises. Both males and females have quite extraordinary noses. The two-tone male has a nose like a bayonet, whilst the female, which is disguised effectively as a twig, has a nose like a thorny club. This genus contains two other equally bizarre species, *L. alluaudi* (see page 129) and *L. pseudoalluaudi*.

*Below* The 'fandrefiala' *(Ithycyphus perineti)*. Though harmless, this arboreal snake is much feared by some rural Malagasy. They believe it can mesmerise people or cattle passing below, then stiffen its body to drop like a spear, tail first, to impale the unfortunate victim. The blood-red body and tail would help create such a myth.

## Family Colubridae

This family, which could be called the 'typical' snakes, is the one to which all other snakes in Madagascar belong.

Quite frequently seen by visitors are the hog-nosed snakes (genus *Leioheterodon*). The largest of the three species, *Leioheterodon madagascariensis*, is reasonably common in rainforests but more likely to be seen in dry forests. It is a handsome yellow and black snake, which can reach a length of 150cm.

## TORTOISES: FAMILY TESTUDINIDAE

Madagascar has the unfortunate distinction of being home to some of the rarest tortoises in the world. Four species are endemic to the island. The largest is the plowshare tortoise or angonoka, *Geochelone yniphora*, (below). Although now restricted to a tiny area around Soalala, south of Majunga (Mahajanga), there is a successful breeding programme at Ampijoroa. This had, by the end of 1995, produced 138 surviving babies, now in various stages of development.

The species gets its name from the long, upturned projection that extends from the plastron (lower shell). Males use this 'plowshare' to joust, trying to lever their opponents on to their backs and gain the attentions of on-looking females. In an unusually robust form of foreplay, they may also use it to roll the female over several times before mating.

The radiated tortoise *(Geochelone radiata)* gets its name from the radiating patterns on its shell. It was formerly abundant but its numbers have been reduced in recent years. It is found in the south and southwest.

The remaining two tortoise species are very much smaller. The dry deciduous forests around Morondava are the only known locality of the flat-tailed tortoise or kapidolo *(Pyxis planicauda)*. The extreme climatic conditions of this area force them to aestivate during the dry season, burying themselves under leaf-litter and sand on the forest floor. Only after substantial rain do they become active. The spider tortoise *(Pyxis arachnoides)* has a much wider range and actually prefers the very dry conditions of the spiny forest areas. Both these species lay a single large egg, producing up to three eggs per season.

Of the four terrapins, only one is endemic. The Madagascar bigheaded turtle, *(Erymnochelys madagascariensis)*, whose nearest relatives are in South America, is found in western lakes and waterways. This protected species has the unfortunate characteristic of growing to an edible size long before reaching sexual maturity. Thus they may be intentionally or accidentally caught in fishing nets before having the chance to reproduce.

*Left* Flat-tailed tortoise or kapidolo *(Pyxis planicauda).* Kirindy (in the wet season) is the best place to see it.

*Below left* Madagascar Radiated tortoise *(Geochelone radiata).*

*Below right* The tortoises of the southwest, though officially protected, often fall prey to protein-hungry local people.

# FROGS

**B**ecause of their permeable skins, amphibians cannot survive in salt water so once Madagascar broke away from Africa those already in residence were marooned. Today, the only amphibians which survive on the island are frogs: there are no newts or salamanders. In splendid isolation these frogs have diversified into the present count of 170 species although the true figure may be nearer 300. New ones are being discovered all the time: from 1990 to 1995 no fewer than 30 new species were described! Only a handful of families are represented (there are no toads, for instance) but some of these show enormous diversity. 99% of Madagascar's frogs are endemic.

## TO FLAUNT OR TO CONCEAL?

The brightly coloured *Mantella* and well-camouflaged *Mantidactylus* genera form a sub-family, Mantellinae, belonging to the 'true frogs' (Ranidae family).

Mantellas show close similarities with the poison arrow frogs (Dendrobatidae) from South America, their colours warning predators to keep away. Their skin secretions are toxic, making them unpalatable. Confident in this protection, they are active by day; most frogs are nocturnal.

In contrast to the gaudy mantellas are the cryptically coloured *Mantidactylus* species. Perhaps none illustrates the art of camouflage better than *Mantidactylus aglavei*, sometimes called the uroplatus frog. Like the leaf-tailed gecko, it exactly matches its preferred resting place – in this case a mossy branch. In conjuction with its colour, frilly projections around its legs complete the disguise. See page 129 for another well-camouflaged *Mantidactylus* species.

*MB*

*Above Mantidactylus pulcher.* These little frogs live only on *Pandanus* palms, concealed by their green colour.

*Top* Painted mantella *(Mantella madagascariensis)*. This colourful species is quite common in the rainforest. During the breeding season they gather in their hundreds by streams and small pools.

*Centre* Golden mantella *(Mantella aurantiaca)*. This tiny frog is only known from areas near Périnet.

*Left* Green-backed mantella *(Mantella laevigata)* lays its eggs in rain water that has collected in tree holes. The South American poison arrow frogs do the same.

# TREE FROGS

Generally speaking, tree frogs have enlarged finger tips to help them grip leaves, and big eyes for night vision. They are usually very photogenic, but to find them you must be prepared to spend many hours at night in the rainforest, systematically homing in on their calls. Rainy nights are the most productive.

One genus to look out for is *Heterixalus* which contains 10 species (see page 19). Most are green or yellow and several are distinctive in having conspicuous stripes running down their flanks. A notable exception is *H. alboguttatus*, one of the larger species (3cm) sometimes seen at Ranomafana. Its basic colour is blue/black with orange spots. However, in sunlight the blue deepens, the spots turn yellow and are ringed in black, while the underside of the hands and feet turn orange.

The 35 known species of *Boophis* are mainly tree frogs which breed in fast-moving streams. Although some are quite drably coloured, many are bright green or rich brown, and they have the most amazingly beautiful eyes! In some species the young change from green to brown as they mature. *Boophis madagascariensis*, for instance, is pale green with black and white markings as a froglet, but matures into a large brown frog that mimics leaf-litter. Its camouflage is enhanced by leaf-like projections from its knees, heels and elbows which break up its outline.

*Below left* Giant, or white-lipped tree frog *(Boophis albilabris)*. For a tree frog this is indeed a giant, reaching nearly 10cm in length. Note the well-developed webbing on the hands and feet. It is found near streams in montane rainforest, including Ranomafana.

*Below right Boophis difficilis*. This little tree frog (about 3cm long) is fairly common in the Périnet area, where males call at night from vegetation 1–2m from the ground. The species name could derive from the frustrations of the taxonomist who finally sorted it out from a host of similar-looking little brown frogs!

*Left* A pair of *Mantidactylus boulengeri* and eggs. The eggs are laid on the forest floor. When the tadpoles hatch they wiggle their way to the nearest water to complete their life cycle.

Some frogs have skipped the free-swimming tadpole stage altogether. Those in the genus *Stumpffia* lay their eggs in foam nests on the forest floor. The tadpoles develop into tiny froglets (only 3mm long) within this foam, eating nothing during the metamorphosis.

*Left* The tomato frog *(Dyscophus antongili)*, is aptly named: that's just what it looks like in colour, shape and size! Apart from a few locations near Tamatave (Toamasina) the tomato frog is only known from the area around the Bay of Antongil (Masoala Peninsula) and the town of Maroansetra. Red means danger in the animal world, and if attacked this frog exudes a sticky white fluid which is toxic and, furthermore, gums up the predator's mouth.

*Left* Not many frog species inhabit the low rainfall areas such as the highlands and the southwest of Madagascar. One exception is *Scaphiophryne gottlebei*, a bizarre frog from Isalo National Park. During the prolonged dry season, it probably conceals itself underground and is only active after heavy rain. This frog is new to science – it was only described in 1992.

*Left Scaphiophryne pustulosa is* found in an equally inhospitable environment, the Ankaratra Mountains in the highlands at about 1,700m, where it breeds in swamps. This frog is known to burrow under the ground.

## The Malagasy and their environment

Periodically Madagascar goes up in flames. Over the centuries the forest that covered much of the island when the first people arrived some 2,000 years ago has been destroyed to make way for the crops and cattle on which the people survive. Folk memories of the great fires that raged across the island about a thousand years ago persist to this day.

For its size Madagascar is not overpopulated. In 1996 about 13 million people lived in an island more than twice the size of Great Britain: a population density of 21 people per square kilometre compared with 228 in Britain. But in 35 years Madagascar's population has doubled and its forest area has halved. Now only about 10% of the original forest remains, and the fertile soil is gone. Much of the once forested highlands are a barren wasteland where nothing will grow. The dry deciduous forests of the west, more vulnerable to fire than the rain-soaked east, are fast disappearing. And in 20 years there will be another 12 million people to feed.

Destruction of habitat is the main threat to Madagascar's wildlife, not hunting. Wild (and endangered) animals are still eaten, but taboos *(fady)* prohibit the killing of many species. As traditional beliefs break down, however, the wildlife is put further at risk. Madagascar's future is in the hands of the Malagasy. But to a hungry man it is the present that matters, not the future.

# INVERTEBRATES

# 6 TO 600 LEGS

Invertebrates (animals without backbones) comprise over 95% of all animal species on earth. Since the break-up of Gondwanaland a myriad of invertebrates have evolved on Madagascar – no one yet knows just how many there are, but it is probably well over 100,000 species. It is beyond the scope of this book to do more than point out some of the most beautiful, colourful or extraordinary of these 'bizarre and wonderful forms'.

An observant visitor will notice an amazing number of these animals in Madagascar's forests. It is always worth turning over stones and logs (though be careful of scorpions), looking carefully on the underside of leaves and examining tree trunks for well-camouflaged invertebrates.

Madagascar's praying mantids are exceptional even in this remarkable family. Many mimic dead leaves, and both nymphs (immature) and adults are totally convincing in their disguise (see pages 128 and 130). Look for them on the shrubs at Périnet and Ranomafana (not on the forest floor). The excellent guides in these reserves will help you in your search. Stick insects are also plentiful and amazing – if you can find them. They mimic

*Below left* Shield bug (order Hemiptera, family Pentatomidae). Some of Madagascar's 200 or so species are quite large and gorgeously coloured, so easy to see.

*Below right* Jewel beetle (order Coleoptera, family Buprestidae). There may be about 20,000 species of beetles in Madagascar, most of which are endemic. The underside of this jewel beetle almost glows with colour.

mossy fronds and thorny green sticks, or look exactly like peeling bark (see page 132).

It is worth spending some time watching insect activity in Madagascar. At Berenty and Ampijoroa, for instance, you will see funnel-shaped holes in the sandy paths. These belong to ant-lion nymphs (order Neuroptera). If you are unkind enough to drop an ant into one of these, the 'lion' lurking below will pounce in a flurry of sand and grab the hapless insect.

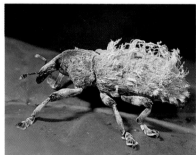

Weevils (order Coleoptera, family Curculionidae).
The countless weevil species (well over 1,000) come in strange forms and colours. The large, hairy ones that live in the eastern rainforest are prized by the local people as food. Roasted, they taste like peanuts.

*Above left* Lixus barbiger from Mananara.

*Above right* Unidentified weevil from Mantady National Park.

*Below* Giraffe-necked weevil *(Trachelophorus giraffa).* Only the males have this extraordinary neck like a mechanical digger. Females are more modestly proportioned. They are very easy to see at Ranomafana, and to a lesser extent at Périnet.

A few of Madagascar's colourful or intriguing insects – and one mite (eight legs, not six).

*Above left* Leaf-mimic praying mantis
*Centre left* Ladybird (family Coccinellidae)
*Below left* Longhorn beetle
    (family Cerambycidae)

*Above right* Giant locust *(Phymateus saxosus)*
*Centre right* Pentatomid bug
*Below right* Mite (family Trombiculidae)

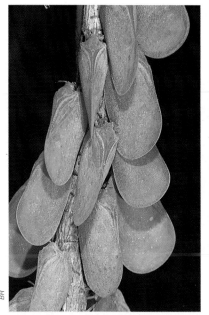

Flatid leaf-bugs. These remarkable bugs *(Phromnia rosea)* are quite easy to see in the wet season. The adults *(above left)* look like pink flowers (and a green form looks just like leaves). The nymphs *(above right and below)* excrete a sort of white waxy substance which 'grows' from the animal like long wispy feathers. If a bird or other predator makes a grab for one of these insects it gets a beakful of white nothing, and the animal hops away (this alone would be sufficient to protect it – it is almost as athletic as a flea). These bugs may be seen at Berenty, but a much bigger and more impressive form is found in the western forests.

The nymphs also excrete a sugary substance which solidifies in lumps on leaves in the forest; this candy treat is eaten by mouse lemurs, especially during the dry season.

*Above* Male hissing cockroach (genus *Gromphadorrhina*). Even the cockroaches in Madagascar are appealing! Although very large (around 7cm) they are relatively slow-moving and very handsome. If prodded males, in particular, emit a loud hiss which both deters predators (presumably) and is used to settle territorial disputes. Hissing cockroaches are easily found on night walks at Berenty where the lighter-coloured genus *Elliptorhina* can be seen waving their antennae out of holes in tree trunks.

*Above* Scorpion *(Grosphus palpator)*. For visitors, scorpions are effectively the only dangerous animals in Madagascar. They are active after rain and at night. During the day they like to hide under stones or in crevasses: beware! Boots or rucksack pockets are favoured places. The sting of the larger species from the western forests is excruciatingly painful to adults and can kill a child.

Millipedes.
Unlike centipedes which bite painfully, millipedes are harmless and attractive. *Above and centre* The pill millipede, (genus *Sphaerotherium*), comes in a variety of colours including green, and rolls itself into an impregnable ball when threatened.

*Above* The handsome giant millipede can exceed 15cm in length. Its many legs (over a hundred but not a thousand, as the name implies) enable it to cover all sorts of terrain – including vertical climbs – at a steady pace. The red colour warns potential predators that it is foul tasting.

# The orchid and the moth

The delicately perfumed, white *Angraecum* orchids of Madagascar are one of the botanical delights of the island. The most famous is the comet orchid, (*Angraecum sesquipedale*).

Some of the British missionaries who worked in Madagascar in the 19th century found the study and recording of the island's natural history a fascinating diversion, especially while travelling through the eastern rainforests en route to the capital from the port of Tamatave. The Reverend William Ellis described the comet orchid in his book *Three Visits to Madagascar during the Years 1853-1854-1856*, published in 1859. He later introduced it into Britain where it was cultivated by James Bateman, a well-known horticulturalist. Mr Bateman sent Charles Darwin some specimens.

In the second edition of his book *The Various Contrivances by which Orchids are Fertilised by Insects* (1904) Darwin wrote: 'The *Angraecum sesquipedale*, of which the large six-rayed flowers, like stars formed of snow-white wax, have excited the admiration of travellers in Madagascar, must not be passed over. A green, whip like nectary of astonishing length hangs down beneath the labellum. In several flowers sent to me by Mr Bateman I found nectaries

eleven and a half inches long, with only the lower inch and a half filled with nectar. What can be the use, it may be asked, of a nectary of such disproportionate length? We shall, I think, see that the fertilisation of the plant depends on this length, and on nectar being contained only within the lower and attenuated extremity. It is, however, surprising that any insect should be able to reach the nectar. Our English sphinxes have proboscides as long as their bodies; but in Madagascar there must be proboscides capable of extension to a length of between ten and eleven inches! This belief of mine has been ridiculed by some entomologists, but we now know from Fritz Muller that there is a sphinx moth in South Brazil which has a proboscis of nearly sufficient length...'

Within a year, a moth with a proboscis of 30cm (12 inches) was found in Madagascar! It was named *Xanthopan morgani praedicta* in honour of Charles Darwin's prediction.

# BUTTERFLIES
# AND MOTHS

*Asota barbonica*

# LAND AND AIR PASSENGERS

Madagascar has around 4,000 species of moth and some 300 butterflies. Moths started to evolve long before butterflies and were undoubtedly part of the fauna of Gondwanaland; butterflies, on the other hand, are more likely to have been dispersed to the island. It is not surprising, therefore, that most families show affinities with African counterparts, but there are some links with India, other parts of Asia, and South America. One of the puzzles is the apparent absence, for instance, of primitive moths such as the Hepialidae family.

There are some surprising absences. The lycaenids (lipenines) of Africa, which feed on lichen, do not occur in Madagascar; their niche is filled with the 300 or so species of footman moths (Lithosiinae).

*Overleaf (page 111)* This *Asota barbonica* moth has just emerged from its chrysalis under a strangler fig, the caterpillar's favourite food. When under threat the moth excretes a frothy substance, pyrazine, which smells strongly of crushed runner-beans. This warns potential predators that the insect is unpalatable. There is a closely related group of moths in South America which also feed on figs.

*Below Cyligramma disturbans* moth (family Noctuidae). This beautiful moth inhabits dark places such as caves and the understorey of the rainforest. The Malagasy believe them to be the embodiment of their ancestors and it is 'fady' (taboo) to kill them.

*Above left* The Urania moth *(Chrysiridia rhipheus)*. At first glance this diurnal moth looks like a butterfly. It is closely related to the genus *Urania* which occurs in South America where it also feeds on Euphorbiaceae plants (genus *Omphalea*). One species of *Chrysiridia* also occurs in Tanzania.

*Above right* Madagascar's largest butterfly, *Atrophaneura anterior*. This swallowtail is a lepidopterological mystery. It appears to be close to the ancestral stock of birdwing butterflies, which poses the question: how did it reach Madagascar? (Butterflies are thought to have evolved after the breakup of Gondwanaland).

*Below* The citrus swallowtail *(Papilio demodocus)*. This is an African species whose caterpillars feed on the leaves of citrus trees. There are, however, three endemic species which are restricted to the western forests.

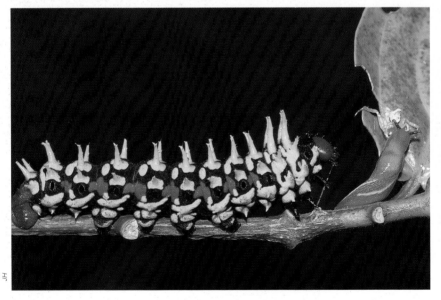

*Above* Comet moth *(Argema mittrei)*. These handsome moths, whose caterpillars spin an equally beautiful silvery cocoon, are much bigger (by a third) than their African relatives. They are bred commercially in Madagascar for collectors.

*Below* Silkworm *(Antherina swaka)*. This is one of Madagascar's endemic silkworms, found in the south feeding on tapia trees. The Malagasy harvest the cocoons for the silk which is used for burial shrouds; the pupae, roasted, make a tasty snack.

# SPIDERS

*Nephila madagascariensis*

# SEX AND THE SINGLE SPIDER

Like all animals, spiders have two major preoccupations: food and sex. However, even a superficial study of the male spider's life (and death) will show that there is little pleasure in the latter. These two drives, eating and reproduction, are in uneasy conflict. Webs are designed to catch insects, and they are made by female spiders. This poses a problem for the male who has to trespass across her dinner table to pair up – she may mistake him for a meal. Since the male spider is often much smaller than the female this confusion is understandable, and he has to go to considerable efforts to achieve his goal without being eaten. His passionless sexual act begins with the depositing of sperm into a tiny homemade silk envelope before transferring the precious fluid into the hollow interior of a specially adapted leg called a pedipalp. Then comes the tricky part – getting it inside the female. Sometimes he taps out a code on the web to announce his presence, or presents the female with a tasty morsel to distract her. Alternatively, like the male golden orb-web spider, he may be so tiny he is not worth bothering about (see *opposite below*).

## Wicked Wives but Marvellous Mums

Following mating the female is transformed into the perfect caring mother. She lays her eggs (many thousand, in some species) on a soft bed of silk which is delicately wrapped into a protective case or cocoon. Depending on the species, this is then either guarded in the web, hidden, or, as with wolf spiders, attached to the spinnerets and dragged about. When the tiny wolf spiderlings hatch they are carried about on their mother's back. During this long period of maternal care, the female does not eat.

*Right* A female wolf spider (family Lycosidae) with her brood. The female's back is covered with special stiff, knobbed hairs which the babies can hold on to.

*Opposite above* Golden orb-web spider. The yellow webs of *Nephila madagascariensis*, stretched between the telegraph wires, are an unmissable feature of many Malagasy towns and one that sets arachnophobes atremble. The silk of these webs is as tough as nylon, and indeed a textile industry using spider silk was attempted towards the end of the last century.

*Opposite below* The male *Nephila* is perched on the female's head and safely out of the way of her mighty jaws. If his pedipalp is loaded he can achieve the sexual act whenever he feels like it: she probably won't even notice!

SG

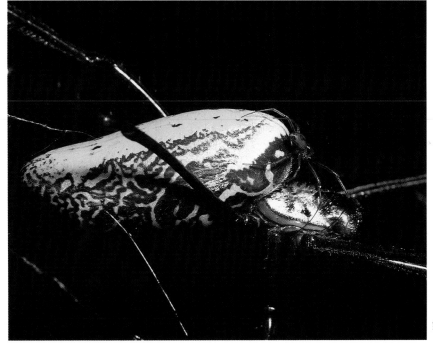

NG

*Above* Green lynx spider *(Peucetia madagascariensis)*, an endemic species found in the eastern forests.

*Right Argiope coquereli*, an endemic species common in the western forests. The conspicuous zig-zag line of web is a strengthening structure called a stabilimentum; it may also attract insects to the web, rather as a bright-coloured flower might.

*Above* Unidentified huntsman spider. As the name implies, these spiders do not use webs to snare they prey, but chase and catch it instead. They have excellent eyesight.

*Below* Thorn spider, subfamily Gasteracanthinae. The thorn spider group is well-known in Africa but has an endemic subspecies in Madagascar.

WL

*Above* Net-throwing spider (family Dinopidae). This is a wide-spread family of spiders with an extraordinarily elaborate – and effective – method of catching their prey. The spider first weaves a rectangular web of various types of silk, most of which are highly elastic. When this 'net' is finished, the spider grasps each corner in its four front legs, cuts the web free from its supports, and hangs head downward with the net at the ready. If an insect passes by, either on the wing or on foot, the spider lunges at it like a butterfly collector, the net being stretchy enough to enmesh even quite large prey.

# SQUASHY THINGS

PH

## SLUGS AND SNAILS, WORMS AND LEECHES

Few visitors are able to enthuse about the island's squashy things but even these are often remarkable looking and have strange lifestyles. They are most commonly encountered after rain. Dry conditions force them to retreat to moist hiding places to avoid dehydration.

Slugs have a certain aesthetic appeal (the creature above appears to be a slug) but it is Madagascar's snails that most interest naturalists because there are so many species – probably close to 400 – and most are endemic. The introduced African giant landsnails (see page 122) indirectly nearly caused the extinction of many of these native species. In 1962 the *Euglandina* snail, a ferocious predator, was introduced into Madagascar in an irresponsible attempt to control the African intruder. Fortunately it did not survive. The *Euglandina* has been responsible for the extinction of all endemic snails in the island of Moorea, in the South Pacific, where it was also introduced.

In addition to the flat worms and leeches, Madagascar has some impressively large earthworms, some nearly 30cm long. These inhabit the eastern rainforests and are a favourite food of tenrecs.

*Above* Giant landsnails mating. These huge snails, whose shells litter some forest floors, were introduced from Africa.

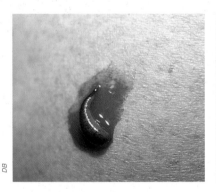

*Above left* Madagascar's leeches are small and do not live in water but hang around on bushes waiting for a host – usually a tourist – to walk past. They attach themselves to their victim with a tail-end sucker and inject an anticoagulant into the wound made with their sharp jaws. The bite is painless but the result is messy. Once the leech has had its fill it falls off, but the bitten area can itch for several days. Leeches are mostly gut: pouches all down the digestive tract enable the animal to hold a large quantity of blood. A satiated leech is about four times its hungry size. At least it does not need to feed again for several months – the time it takes to digest this banquet of blood.

*Above right* This impressive animal is not a leech but a flatworm. Madagascar has several species, many of which are brightly coloured with two-tone stripes running down their bodies. All have hammer-shaped heads and are harmless.

# MADAGASCAR AT NIGHT

Madagascar scops owl *(Otus rutilus)*

# FROM DUSK TO DAWN

One of the joys of walking in a Malagasy forest is the safety. There are no large animals hiding in the trees, no venomous snakes concealed in the foliage and almost no horrible creepy-crawlies lurking in the undergrowth (scorpions are the exception). To add to the excitement, the day shifts and night shifts are completely different. Come dusk, when all the familiar daytime creatures are bedding themselves down, a new cast of characters is stirring. These include nocturnal lemurs, carnivores, nocturnal birds, *Uroplatus* geckos, frogs and a myriad of insects. These are Madagascar's spirits of the night.

Of course, the fundamental problem to seeing nocturnal animals is the dark. However, there are tell-tale signs to look for. The back of the eye (retina) in most nocturnal animals is highly reflective. This specialised area, the tapetum, increases the sensitivity of the eye to dim light. It also reflects incoming light straight back out again, so the eyes of an animal looking towards you appear to glow red in the torchlight. This is commonly called 'eyeshine'.

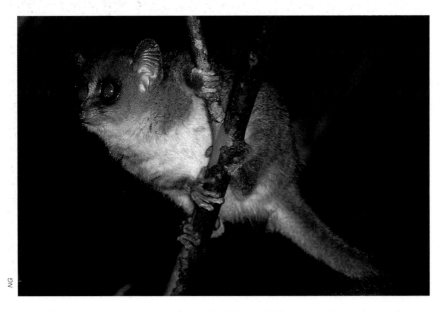

NG

*Above* Brown mouse lemur *(Microcebus rufus)*. Many of Madagascar's most interesting animals may only be seen at night. Guided night walks are available in many of the national parks and reserves and can be very rewarding (see page 126). Some of the guides put out food to attract certain animals: mouse lemurs come for bananas, fosas are lured by the smell of grilled fish, and striped civets by raw or cooked chicken.

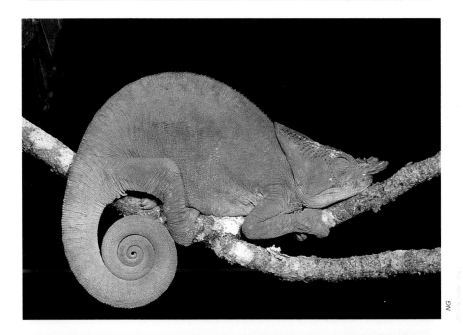

*Above* Chameleons are easier to see in the dark because they turn a much lighter shade; some species become almost white. The Parson's chameleon above is in typical sleeping posture in the top picture: tail coiled, chin resting on the branch. Woken by the flash of the first photo, his skin has already darkened and his tail uncoils in preparation for a quick getaway.

# A shot in the dark

Night time wildlife watching (and photography) is not just a case of where to look but how to look. Here are some hints.

● Avoid times of the full moon. Many animals will be in hiding.

● Wear appropriate clothing: subdued colours and fabrics that do not rustle.

● Bring binoculars. They are helpful even in torchlight.

● Use a head torch in conjunction with a powerful hand torch. Halogen bulbs are best. Bring plenty of new batteries and replacement bulbs.

● Macro photography is very successful at night. See page 132.

Now that you are ready, what do you look for? At first only use the head torch and set the beam focus so that it throws a fairly wide and diffuse pool of light. Then search the pool of light using your peripheral vision, that is, avoid looking directly at something of interest. In dim light you will actually see it more clearly out of the corner of your eye.

If something catches your attention – a pair of red glowing eyes, for instance – illuminate it fully with the hand torch and check it out using binoculars. Should the animal take flight, shut down the lights, wait and start again – it may resettle close by.

Here's where to look for your favourite creatures:

**Lemurs** The smaller species can be seen anywhere, but it is easiest to concentrate on the lower, thinner branches that they prefer. Lepilemurs and woolly lemurs are generally found clinging to narrower vertical trunks. Always investigate the areas around tree holes, especially at dusk when animals have just woken up.

**Chameleons** Often sleep at the ends of branches, and their nighttime light colour stands out in a torch beam. Always check the lower branches around the outside of bushes and trees. Stump-tailed chameleons are found really low down or on the ground. It's just a question of grubbing about.

**Uroplatus** The larger species like *U. fimbriatus* hunt from ground level to the canopy. Their eyes do shine red, but not as brightly as those of lemurs, and you'll only see one eye at a time. The smaller species tend to be on finer branches of undergrowth and amongst leaves, so are much harder to find.

**Frogs** Many species tend to be very vocal at night, so are heard but not always seen. They often sound closer than they actually are.

**Invertebrates** Some also have eyeshine. Check the forest floor and tree bark for spiders, the ends of twigs where butterflies and moths sometimes roost, the underside of leaves for stick insects and mantids, and fallen trees and rotten logs for millipedes, beetles and bugs.

**Birds** Go out a couple of hours before dawn. You can then catch the nocturnal birds and be on the spot at daybreak when the diurnal species are at their most active

# CAMOUFLAGE

Collared nightjar *(Caprimulgus enarratus)*

# SAFETY MATCHES

In the absence of large mammals in Madagascar, many of the small animals – mainly reptiles, frogs and invertebrates – that make up a major part of the diet for most of the island's carnivores hide to escape their predators. Some have evolved the ability to merge into their background to a remarkable degree.

The collared nightjar (page 127) is an excellent example of cryptic colouration. This nocturnal bird generally roosts on the ground during the day; the brown and yellow patterns of its plumage so exactly matches the background of leaves that it is very difficult to see. The related frogmouths of Australia are similarly disguised. Like all well-camouflaged animals, nightjars keep still when in danger, only flying away when the threat gets uncomfortably close.

Cryptic colouring ensures that an animal blends with its background. Others have evolved bodies that look just like a part of the forest: leaves or twigs. Reptiles such as the *Uroplatus* gecko or *Brookesia* chameleon specialise in mimicking dead leaves, and, of course, stick insects look just like twigs.

*Opposite* The Parson's chameleon and *Mantydactylus* frog are both examples of cryptic colouring. Their patterns and colours blend with the leaves (green and brown) of their forest habit.

*Below* Twig-mimic mantis. Many insects and some species of snakes and frogs look like twigs, dead leaves, or moss. They do more than blend into the background – they are part of it.

GT

NG

*Opposite: upper left* Moss-mimic stick insect (*Parectatosoma macquersi*)

*Opposite: upper right* Stick insect

*Opposite: centre* Leaf-mimic praying mantis (*Phyllocrania mudens*)

*Opposite: lower left* Lichen-mimic insect, probably *Flatoides* sp.

*Opposite: lower centre* Cryptically coloured nymphs of the bug *Libyaspis coccinelloides*

*Opposite: lower right* Cryptically coloured *Panogena jasmini* hawkmoth

*Above left* Female twig-mimic snake (*Langaha alluaudi*)

*Above right* Leaf-tailed gecko (*Uroplatus fimbriatus*)

*Above* Stump-tailed chameleon (*Brookesia stumpffi*)

# Photography

Much of the wildlife in Madagascar is approachable. To take excellent photographs like those in this book you do not need very expensive equipment; the secret is getting into the right place at the right time and knowing what to do when you get there. Here are some hints.

● The camera body should have the option of a wide range of different lenses and other accessories such as flashguns.

● Choose the best lens you can afford. The type of lens will be dictated by the subjects you most wish to photograph.

● 'Professional' quality film is worth the extra expense. Recommended are Fujicolour Reala 100 for prints, and Fujichrome Velvia 50 and Fujichrome Provia 100 or Sensia 100 for transparencies.

● Whenever possible use a tripod with a cable release. Without a tripod a good rule of thumb is never use a shutter speed slower than the reciprocal of the focal length of the lens, ie: with a 50mm lens, 1/60th second, with 100–135mm lenses, 1/125th second.

● Consider the background to your photographs. Getting this right is often the difference between a mediocre and a memorable image.

● If it does not look good through the viewfinder, it will never look good as a picture. Be patient. Wait until the background is uncluttered and there is a natural highlight in the eyes. It makes all the difference.

**Birds and Mammals**
A lens of at least 300mm is necessary for bird photography to produce a reasonable image size. Optics of this length should always be held steady on a tripod. The subject needs to be well-lit, but the majority of species in Madagascar are forest dwellers; a powerful flashgun adds extra light to pep up the subject. For use with longer lenses a flash unit with a guide number of 40 or more is best. This will allow subjects over 10m away to be photographed, and the extra light can transform an otherwise dreary picture by 'lifting' the subject out of its surroundings and putting the all important highlights into the eyes.

**Small Things: Macro Photography**
Macro lenses of around 50mm and 100mm cover most subjects and these create images up to half life size in the viewfinder. For further enlargement extension tubes will be required.

At distances of less than a metre, apertures as small as between f16 and f32 are necessary to ensure adequate depth of field, plus flash to provide enough light. Most manufacturers make macro-flash set ups including ring-flashes. These, however, tend to produce very 'flat' lighting that can make the subject look two dimensional. A better solution is to build your own bracket and use one or two small flashguns to illuminate the subject from the side. The resulting shadows give the subject depth.

# Further Reading

## Travel and other natural history guides

*Guide to Madagascar.* Hilary Bradt. Bradt Publications. A thorough guide to the country that is regularly updated.

*Globetrotter Travel Guide to Madagascar.* Derek Schuurman (1996). Struik/New Holland. A travel guide aimed at the ecotourist.

*Madagascar: A Natural History.* Ken Preston-Mafham (1991). Facts on File. A large format coffee table book.

*Lemurs of Madagascar – a Tropical Fieldguide.* Russell Mittermeier et al (1994). Conservation International. A pocket reference to the lemurs.

*A Guide to the Birds of Madagascar.* Oliver Langrand (1990). Yale UP. Ideal for the keen birdwatcher.

*A Fieldguide to the Amphibians and Reptiles of Madagascar.* Frank Glaw and Miguel Vences (1994, 2nd ed). A review of the island's amazing herpetofauna.

## Specialist natural history

*Lemurs of Madagascar: An Action Plan for their Conservation.* Russell Mittermeier et al (1992). The latest priorities for the conservation of lemurs.

*Lemurs of Madagascar and the Comoros – UICN Red Data Book.* Caroline Harcourt and Jane Thornback (1990). UICN – Cambridge and Switzerland. A review of the conservation status of lemurs.

*Madagascar: Revue de la Conservation et des Aires Protegees.* Martin Nicoll and Olivier Langrand (1989). WWF – Switzerland. A comprehensive look at the island's protected areas. Mainly in French.

*Madagascar: An Environmental Profile.* Alison Jolly et al (1984). Pergamon.

*Chameleons – Nature's Masters of Disguise.* James Martin and Art Wolfe (1992). Blandford Press. A fully illustrated look at the world of chameleons.

## Background reading

*Madagascar: A World Out of Time.* Frans Lanting (1990). Hale. A photographic essay of the people, wildlife and landscapes of Madagascar.

*The Aye-aye and I.* Gerald Durrell (1992). Harper Collins. Gerald Durrell recounts his last collecting expedition to Madagascar.

*Lemurs of the Lost World.* Jane Wilson (1995, 2nd ed). Impact Books. An account of the author's expedition to the Ankarana caves in northern Madagascar.

# Index

Panther chameleon

# Index of Scientific Names

# MADAGASCAR WILDLIFE